The ESSENTIALS of TRANSPORT PHENOMENA II

Staff of Research and Education Association,
Dr. M. Fogiel, Director

This book is a continuation of *"THE ESSENTIALS OF TRANSPORT PHENOMENA I"* and begins with Chapter 8. It covers the usual course outline of Transport Phenomena II. Earlier/basic topics are covered in *"THE ESSENTIALS OF TRANSPORT PHENOMA I"*.

REA

Research and Education Association
505 Eighth Avenue
New York, N.Y. 10018

THE ESSENTIALS OF TRANSPORT PHENOMENA II

Printed in the United States of America

Library of Congress Catalog Card Number 87-61803

International Standard Book Number 0-87891-629-6

WHAT "THE ESSENTIALS" WILL DO FOR YOU

This book is a review and study guide. It is comprehensive and it is concise.

It helps in preparing for exams, in doing homework, and remains a handy reference source at all times.

It condenses the vast amount of detail characteristic of the subject matter and summarizes the **essentials** of the field.

It will thus save hours of study and preparation time.

The book provides quick access to the important facts, principles, theorems, concepts, and equations of the field.

Materials needed for exams, can be reviewed in summary form – eliminating the need to read and re-read many pages of textbook and class notes. The summaries will even tend to bring detail to mind that had been previously read or noted.

This "ESSENTIALS" book has been carefully prepared by educators and professionals and was subsequently reviewed by another group of editors to assure accuracy and maximum usefulness.

Dr. Max Fogiel
Program Director

CONTENTS

This book is a continuation of *"THE ESSENTIALS OF TRANSPORT PHENOMENA I"* and begins with Chapter 8. It covers the usual course outline of Transport Phenomena II. Earlier/basic topics are covered in *"THE ESSENTIALS OF TRANSPORT PHENOMA I"*.

Chapter No. **Page No.**

CHAPTER 8

FORCED CONVECTION

8.1 DEFINITION

(1) $$t_b = \frac{\int_0^R 2\pi r t \rho C_p u dr}{\int_0^R 2\pi r \rho C_p u dr} = \text{bulk temperature}$$

(2) $$t_m = \frac{t_s + t_f}{2} = \text{mean temperature}$$

where t_s = surface temperature

t_f = free stream fluid temperature

(3) $$t_{mf} = \frac{t_s + t_b}{2} = \text{mean bulk temperature}$$

(4) $$t_{mb} = \frac{t_{i,b} + t_{o,b}}{2} = \text{mean bulk temperature}$$

where $t_{i,b}$ = inlet bulk temperature

$t_{o,b}$ = outlet bulk temperature.

(5) $NU_x = \frac{hx}{k}$ = local Nusselt number

(6) $NU_L = \frac{hL}{k}$ = average Nusselt number

(7) $R_{ex,c} = \frac{UX_c}{\nu}$ = critical Reynolds number, X_c is the distance at which transition takes place.

8.2 FLOW OVER A FLAT PLATE

8.2.1 LAMINAR FLOW

$$NU_X = 0.332\, R_{ex}^{\frac{1}{2}}\, Pr^{\frac{1}{3}}$$

$$NU_L = 0.664\, R_{eL}^{\frac{1}{2}}\, Pr^{\frac{1}{3}}$$

Restrictions:

(1) $0.6 \leq Pr \leq 50$

(2) $100 < R_e < R_{ex,c} \simeq 5 \times 10^5$

(3) Evaluate properties at t_m

(4) Dissipation is neglected

$$NU_X = 0.565(R_{ex} \cdot Pr)^{\frac{1}{2}}$$

$$NU_L = 1.130(R_{eL} \cdot Pr)^{\frac{1}{2}}$$

Restrictions:

(1) $Pr < 0.05$

(2) $R_e < R_{ex,c} \cong 5 \times 10^5$

(3) Evaluate properties at t_m

(4) Dissipation is neglected

8.2.2 TURBULENT FLOW

$$NU_X = 0.0296\, R_{ex}^{0.8} \cdot Pr^{\frac{1}{3}} \qquad 5 \times 10^5 < R_{ex} < 10^7$$

$$NU_X = 0.185\, R_{ex} (\log R_{ex})^{-2.584} \cdot Pr^{\frac{1}{3}} \qquad R_{ex} > 10^7$$

Restrictions:

(1) $0.6 \leq Pr \leq 60$

(2) Properties at t_m

(3) Dissipation is neglected

$$NU_L = \left[0.037 R_{eL}^{0.8} - \frac{A}{2}\right] Pr^{\frac{1}{3}} \qquad 5 \times 10^5 < R_{eL} < 10^7$$

$$NU_L = \left[0.228 R_{eL} (\log R_{eL})^{-2.584} - \frac{A}{2}\right] Pr^{\frac{1}{3}} \qquad 10^7 < R_{eL} < 10^9$$

Restrictions:

(1) $0.6 \leq Pr \leq 60$

(2) $R_{ex,c} = 5 \times 10^5$

(3) Evaluate properties at t_m

(4) Dissipation is neglected

(5) Obtain A from table 8.1

Table 8.1	
Transition Reynolds Number $R_{ex,c}$	A
3×10^5	1055
5×10^5	1743
1×10^6	3341
3×10^6	8944
Laminar portion is ignored	0

8.3 FLOW INSIDE CYLINDRICAL PIPES AND TUBES

8.3.1 LAMINAR FLOW

$$NU_D = \frac{h \cdot D}{K} = 3.66$$

Restrictions:

(1) Fully developed flow

(2) $R_{eD} < 2300$

(3) Constant surface temperature, t_s = constant

(4) Evaluate properties at t_{mb}

(5) $\left(\frac{D}{L}\right) \cdot R_{eD} \cdot Pr < 10$

$$NU_D = 1.86\left[\left(\frac{D}{L}\right) R_{eD} \cdot Pr\right]^{\frac{1}{3}} \left(\frac{\mu}{\mu_s}\right)^{0.14}$$

Restrictions:

(1) $0.48 < Pr < 16{,}700$

(2) Evaluate properties except μ_s at t_{mb}

Evaluate μ_s at t_s

(3) $\left(\frac{D}{L}\right) R_{eD} \cdot Pr > 10$

8.3.2 TURBULENT FLOW

$$NU_\infty = 0.023 R_{eD}^{0.8} \; Pr^n$$

Restrictions:

(1) $n = 0.4$ for $t_s > t_b$

(2) $n = 0.3$ for $t_s < t_b$

(3) $0.7 < Pr < 160$

$10^4 < R_{eD} < 10^6$

(4) $|t_s - t_b| < 6°C$ for liquids

$|t_s - t_b| < 60°C$ for gases

(5) Evaluate properties at t_b

(6) Fully developed flow

Effect of starting length is negligible

$$\frac{L}{D} \geq 60$$

(7) $R_{eD} > 2300$

$$NU_{\infty} = 0.027 R_{eD}^{0.8}\ Pr^{\frac{1}{3}} \left(\frac{\mu}{\mu_s}\right)^{0.14}$$

Restrictions:

(1) $0.7 < Pr < 160$

(2) $10^4 < R_{eD} < 10^6$

(3) Evaluate properties except μ_s at t_b

Evaluate μ_s at t_s

(4) Fully developed flow

(5) Effect of starting length is negligible $\frac{L}{D} \geq 60$

8.4 HEAT-TRANSFER COEFFICIENT FOR TRANSITION FLOW INSIDE A PIPE

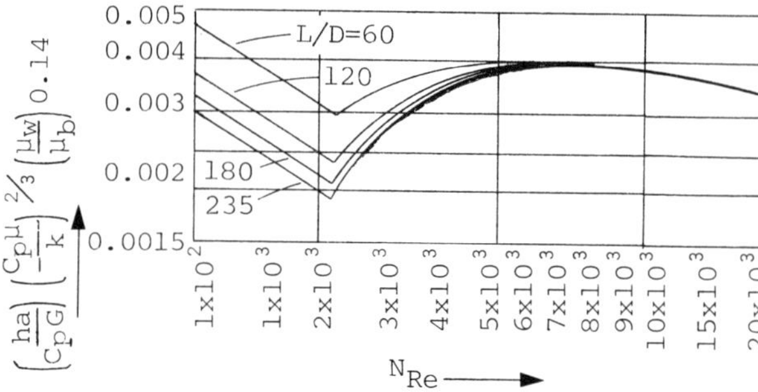

Fig. 8.1

Figure 8.1 shows the relationship between the various heat transfer parameters and the Reynolds number between 2100 and 10^4.

8.5 FLOW NORMAL TO A CIRCULAR CYLINDER

Table 8.2

Range of R_{eD}	C	m
1 - 40	0.75	0.4
$4 \times 10 - 1 \times 10^3$	0.51	0.5
$1 \times 10^3 - 2 \times 10^5$	0.26	0.6
$2 \times 10^5 - 1 \times 10^6$	0.076	0.7

$$NU_\infty = C\, R_{eD}^{m}\, Pr^{n} \left(\frac{Pr_\infty}{Pr_s}\right)^{\frac{1}{4}}$$

Restrictions

(1) $0.7 < Pr < 500$

(2) $1 < R_{eD} < 10^6$

(3) $n = 0.37 \quad Pr \leq 10$

$n = 0.36 \quad Pr > 10$

(4) Evaluate properties at $t_m = \frac{t_\infty + t_s}{2}$

except Pr_∞ , Pr_s

Pr_∞ at t_∞

Pr_s at t_s

(5) Single tube or cylinder

(6) Obtain C and m from table 8.2

8.6 FLOW NORMAL TO NON-CIRCULAR CYLINDER

Table 8.3

Geometric Configuration	Re_{Df}	C	η
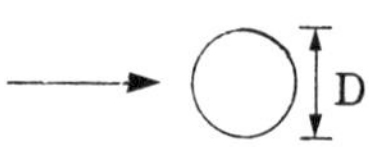	0.4 to 4 4 to 40 40 to 4,000 4,000 to 40,000 40,000 to 400,000	0.989 0.911 0.683 0.193 0.0266	0.330 0.385 0.466 0.618 0.805
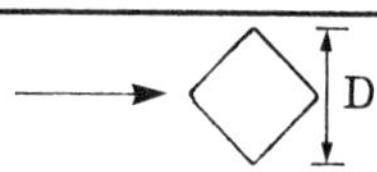	2,500 to 7,500 5,000 to 100,000	0.261 0.222	0.624 0.588
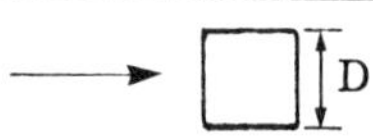	2,500 to 8,000 5,000 to 100,000	0.160 0.092	0.699 0.675
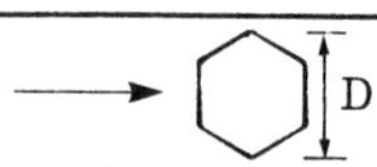	5,000 to 19,500 19,500 to 100,000	0.144 0.035	0.638 0.782
	4,000 to 15,000	0.205	0.731
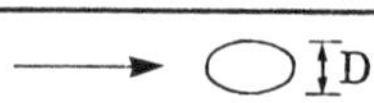	2,500 to 15,000	0.224	0.612
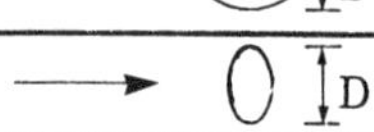	3,000 to 15,000	0.085	0.804
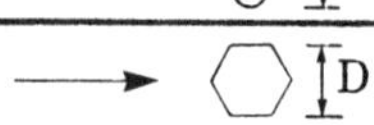	5,000 to 100,000	0.138	0.638

$$NU_D = C\,R_{e\infty}^{n}\,Pr^{\frac{1}{3}}$$

Restrictions

(1) Correlation applicable to gases

(2) Evaluate properties at t_m

(3) Obtain C and m from table 8.3

8.7 FLOW OVER BANKS OF TUBES

$$R_{eD,m} = \frac{U_m \cdot D}{\nu}$$

$$U_m = U_\infty \frac{S_T}{S_T - D} \qquad U_m = \begin{cases} U_\infty \dfrac{S_T}{S_T - D} & \text{for } 2(S_D - D) > S_T - D \\[2ex] U_\infty \dfrac{S}{2(S_T - D)} & \text{for } 2(S_D - D) < S_T - D \end{cases}$$

U_m = maximum velocity

S_T = Centers of the tube normal to the flow

S_L = Tube centers parallel to the flow

$$\boxed{NU_D = CR_{eD,m}^{n}\,Pr^{0.36}\left(\frac{Pr_\infty}{Pr_s}\right)^{\frac{1}{4}}}$$

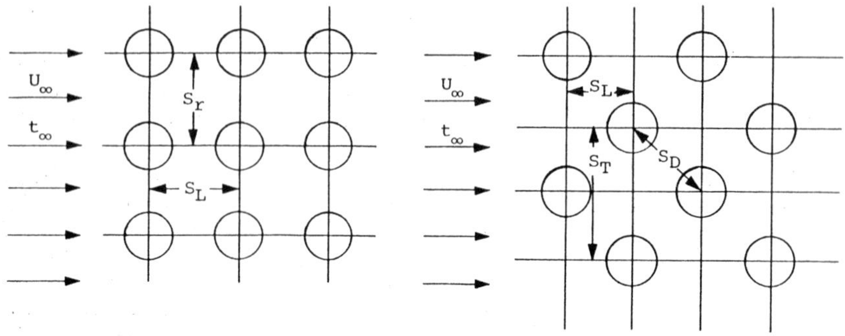

$$S_D = \left[S_L^2 + \left\{ \frac{S_T}{2} \right\}^2 \right]^{1/2}$$

ALIGNED ARRANGEMENT

STAGGERED ARRANGEMENT

Fig. 8.2

Restrictions:

(1) $0.7 < Pr < 500$

(2) $10 < R_{eD,m} < 10^6$

(3) Evaluate properties at t_m except Pr_∞, Pr_s

Pr_∞ at t_∞

Pr_s at t_s

(4) Obtain C and n from table 8.4

(5) Correlation is valid for 20 or more rows

For less than 20 rows use the correction factor:

$$\frac{(NU_D)_N \text{ rows}}{(NU_D)_{20} \text{ rows}}$$

Table 8.4				
R_{eDm} Range	Aligned		Staggered	
	C	η	C	η
$10 - 10^2$	0.8	0.4	0.9	0.4
$10^2 - 1 \times 10^3$	Use Zhukauskas correlation for a single tube (Section C)			
$1 \times 10^3 - 2 \times 10^5$	$\frac{S_T}{S_L} < 0.7$: Heat transfer is very poor and this design is not recommended		$\frac{S_T}{S_L} < z$:	
			$0.35\left(\frac{S_T}{S_L}\right)^{\frac{1}{5}}$	0.6
	$\frac{S_T}{S_L} > 0.7$:			
	0.27	0.63	0.40	0.60
$2 \times 10^5 - 1 \times 10^6$	0.021	0.84	0.022	0.84

8.8 FLOW PAST SINGLE SPHERE

When a sphere is being heated or cooled by a fluid flowing past it, the following equation can be used to predict the average heat-transfer coefficient.

$$N_u = 2.0 + 0.60\, R_e^{0.5}\, Pr^{\frac{1}{3}}$$

Restrictions:

(1) $1 \leq R_e \leq 70{,}000$

(2) $0.6 \leq Pr \leq 400$

8.8.1 HEAT TRANSFER FOR FLOW IN PACKED BEDS

For the heat transfer of gases in beds of sphere and Reynold's number ranges from 10 to 10,000

$$\varepsilon T = \varepsilon \frac{h}{cv'\rho}\left(\frac{C_p \mu}{K}\right)_f^{\frac{2}{3}} = \frac{2.876}{NR_e} + \frac{0.3023}{NR_e^{0.35}}$$

v' = Superficial velocity based on the area of the empty container

ε = Void fraction

$NR_e = D\,G'/\mu_f$

$G' = v'\rho$

Subscript 'f' indicates properties evaluated at film temperature with other being taken at bulk temperature.

8.9 LOCAL OVERALL COEFFICIENT OF HEAT TRANSFER

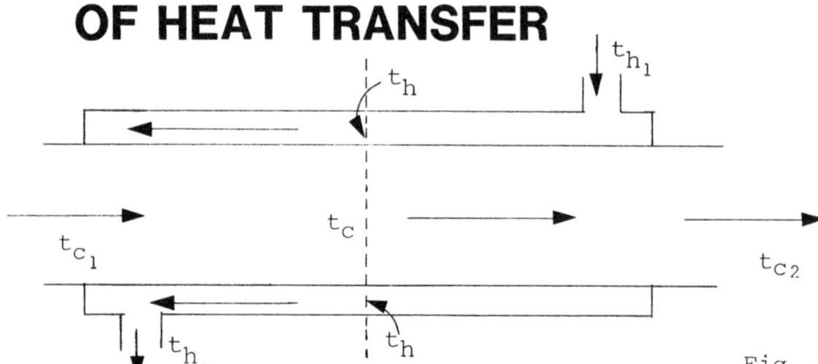

Fig. 8.3

Consider a counter-flow heat exchanger as shown in Figure 8.4. The rate of heat transfer per unit area is generally given by

$$\boxed{dq = U dA \, \Delta t_0} \tag{8.1}$$

U = Local overall coefficient of heat transfer

dA = Heat transfer surface area

Δt_0 = Average effective temperature difference for the entire heat exchanger.

For example, the overall heat transfer coefficient U for a plane wall is given by

$$\boxed{U = \frac{1}{\frac{1}{h_o} + \frac{L}{K} + \frac{1}{h_i}}} \tag{8.2}$$

where h_o and h_i are local heat transfer coefficients for the outer and inner surface of the wall. For the counter-flow heat exchanger

$$\boxed{U_i = \frac{1}{\frac{1}{h_i} + \frac{A_i \ln\left(\frac{r_o}{r_i}\right)}{2\pi k L} + \frac{A_i}{A_o}\frac{1}{h_o}}} \tag{8.3}$$

$$\boxed{U_o = \frac{1}{\frac{A_o}{A_i}\frac{1}{h_i} + \frac{A \ln\left(\frac{r_o}{r_i}\right)}{2\pi k L} + \frac{1}{h_o}}} \tag{8.4}$$

8.10 HEAT TRANSFER THROUGH SOLID CONDUCTOR

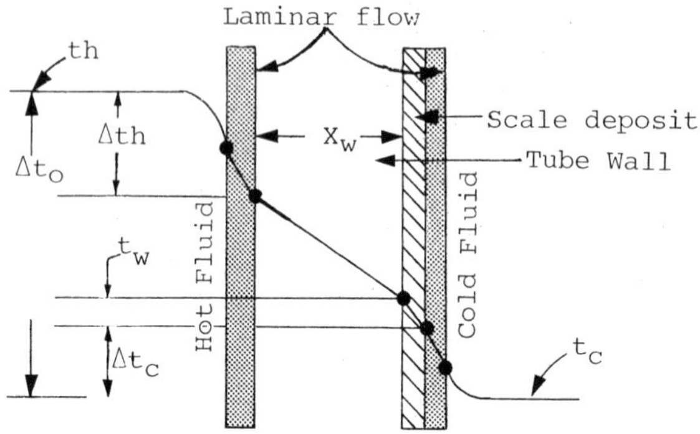

Fig. 8.4

The rate of heat transfer by conduction through a tube wall of thickness X_w is given by

$$dq = k_w \, dA_w \, \frac{\Delta t_w}{X_w} \qquad (8.5)$$

where dA_w represents logarithmic mean area. Since the rate of conduction is the same flow of heat despite the scales of thickness X_d;

Then,

$$dq = \frac{k_d \, dA_c \, \Delta t_d}{X_d} = h_d \, dA_c \, \Delta t_d$$

where dA_c = Cooling surface area

$$h_d = \frac{k_d}{X_d}$$

The relationship accounting for the difference in temperature of the outer surface of the deposit and bulk temperature of the cold fluid Δt_c is

$$dq = h_c \, dA_c \, \Delta t_c \qquad (8.6)$$

8.11 HEAT EXCHANGERS

Three types of heat exchanger are in use:

(1) Double pipe heat exchanger

(2) Shell and tube heat exchanger

(3) Cross flow heat exchanger.

8.11.1 DOUBLE PIPE HEAT EXCHANGER

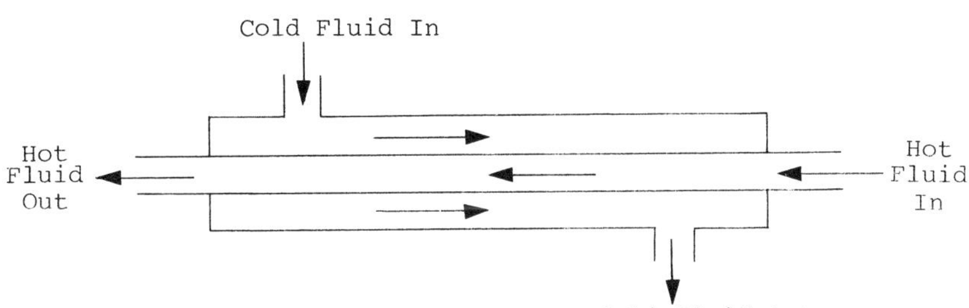

Fig. 8.5

The simplest type of heat exchanger is the double pipe heat exchanger. This type of heat exchanger is useful where small flow rates are involved. The fluid to be cooled flows in the inner pipe while the cooling fluid flows in the annular space. The direction of the flow may be co-current or counter-current.

8.11.2 SHELL AND TUBE EXCHANGER

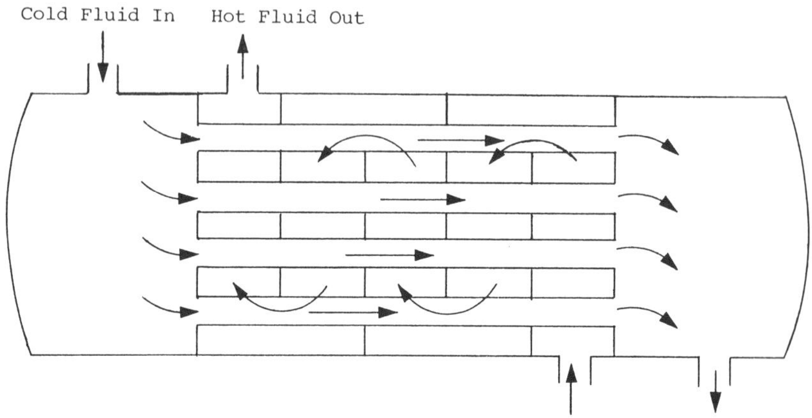

Fig. 8.6

If larger flow rates are involved a shell and tube exchanger may be used. In this type of exchanger the flows are continuous. The exchanger shown in the figure is known as 1-shell-pass and 1-tube-pass. Cross baffles are used to increase the velocity of the fluid. This adds turbulence which increases the shell side heat transfer.

8.11.3 CROSS FLOW EXCHANGER

This type of exchanger is used when a gas is to be heated or cooled. The fluid which is liquid, flows inside the tubes and gas flows across the tube boundle by forced convection.

8.12 LOG MEAN TEMPERATURE DIFFERENCE

CASE I: Constant U, Counter flow operation

$$\dot{q} = \dot{M}_g C_g [T_{g\,out} - T_{g\,in}] = UA\Delta t_{lm}$$

$$= M_l C_l [T_{l\,out} - T_{l\,in}]$$

$$\Delta t_{lm} = \frac{\Delta T_2 - \Delta T_1}{\ln \frac{\Delta T_2}{\Delta T_1}} = \frac{(T_1 - T_2') - (T_1' - T_2)}{\ln \left(\frac{T_1 - T_2'}{T_1' - T_2} \right)}$$

Counter Flow

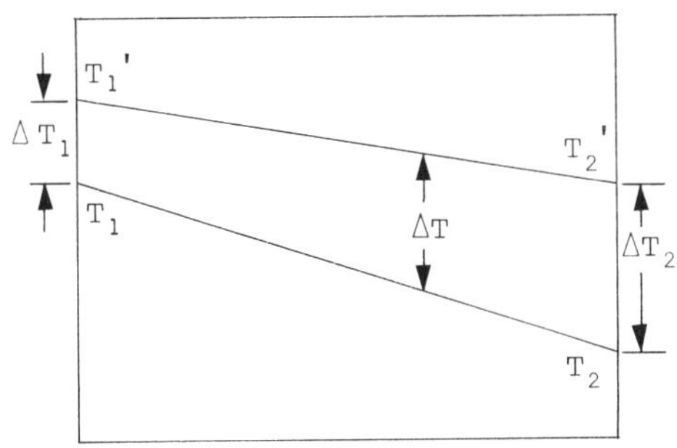

Fig. 8.7

T_1' = Hot fluid temperature at the inlet

T_2' = Cold fluid temperature at the inlet

T_1 = Hot fluid temperature at the outlet

T_2 = Cold fluid temperature at the outlet

ΔT_2 = Temperature difference between bulk and cooling fluid

ΔT_1 = Temperature difference between bulk and hot fluid

C_l = Heat capacity of liquid

$\dot{M}_l$ = Mass flow rate of liquid

$\dot{M}_g$ = Mass flow rate of hot gas

C_g = Heat capacity of hot gas

CASE II: Parallel flow or co-current flow

$$\Delta t_{lm} = \frac{(T_{h_2} - T_{c_2}) - (T_{h_1} - T_{c_1})}{\ln[(T_{h_2} - T_{c_2})/(T_{h_1} - T_{c_1})]}$$

T_{h_2} = Temperature of the hot fluid at the outlet

T_{c_2} = Temperature of the cold fluid at the outlet

T_{h_1} = Temperature of the hot fluid at the inlet

T_{c_1} = Temperature of the cold fluid at the inlet

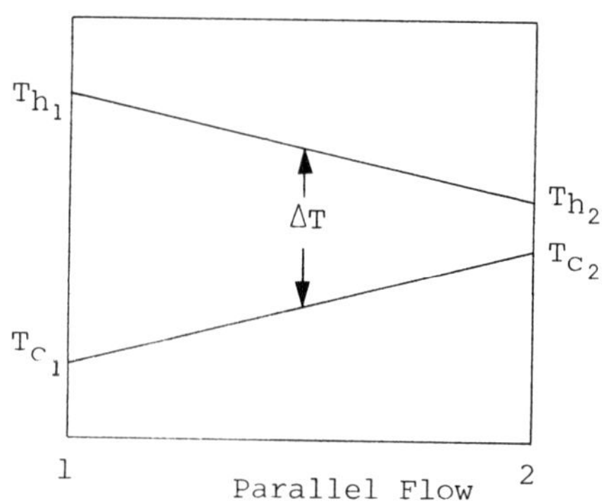

Fig. 8.8

CASE III: Constant U, Multipass flow:

$$\dot{q} = U_i A_i \ \Delta T_m = U_o A_o \ \Delta T_m$$

$$\Delta T_m = F_T \ \Delta T_{lm}$$

where F_T is the correction factor.

$$\Delta T_{lm} = \frac{(T_{hi} - T_{co}) - (T_{ho} - T_{ci})}{\ln[T_{hi} - T_{co}/T_{ho} - T_{ci}]}$$

F_T can be obtained from the plot of

$$P = \frac{T_{co} - T_{ci}}{T_{hi} - T_{ci}}$$

against F_T for different values of

$$R = \frac{T_{hi} - T_{ho}}{T_{co} - T_{ci}}$$

where

$T_{c,o}$ = Temperature of cold fluid at outlet

$T_{c,i}$ = Temperature of cold fluid at inlet

$T_{h,o}$ = Temperature of hot fluid at outlet

$T_{h,i}$ = Temperature of hot fluid at inlet

$\dot{q}$ = Heat per unit time

R = Ratio of the temperature difference of the hot to that of cold fluid in a heat exchanger.

CASE IV: Variable U, Parallel or Counter-Flow Exchanger

If the overall heat transfer coefficient varies with temperature, the exchanger may be considered to be divided into a number of small exchangers in series. We assume that in each small exchanger U is linear with temperature hence

$$U = a + b(\Delta t_o)$$

$$\boxed{\dot{q} = A\,\frac{U_1(\Delta t_o)_2 - U_2(\Delta t_o)_1}{\ln[U_1(\Delta t_o)_1 / U_2(\Delta t_o)_2]}}$$

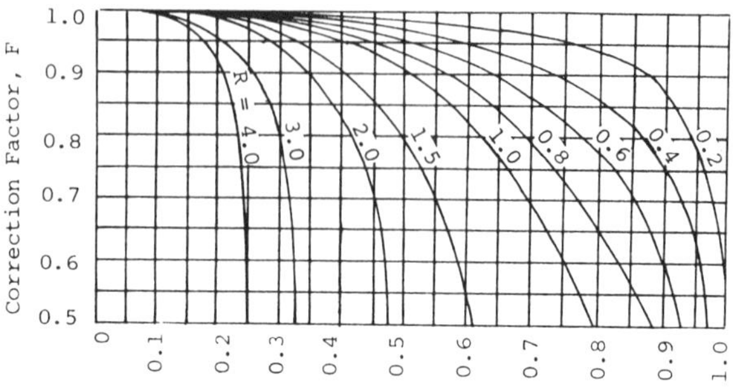

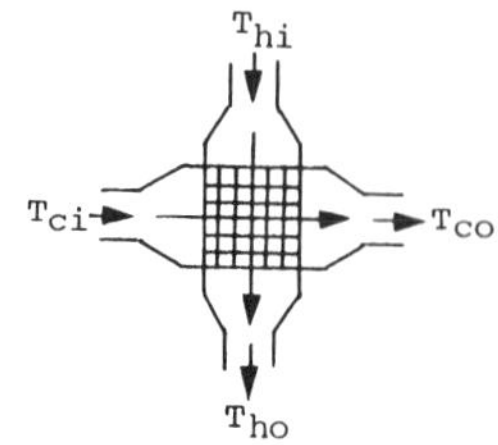

$$R = \left(\frac{T_{hi} - T_{ho}}{T_{co} - T_{ci}}\right)$$

$$P = \left(\frac{T_{co} - T_{ci}}{T_{hi} - T_{ho}}\right)$$

Fig. 8.9 Correction factor for- Crossflow 'LMTD'

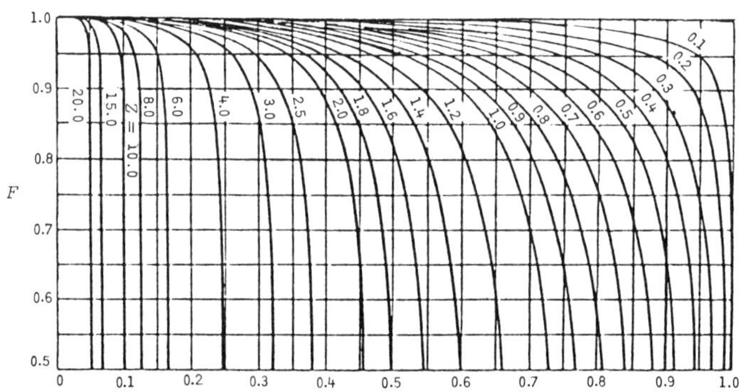

$$P = \left(T_{t\ out} - T_{t\ in}\right) / \left(T_{s\ in} - T_{t\ in}\right)$$

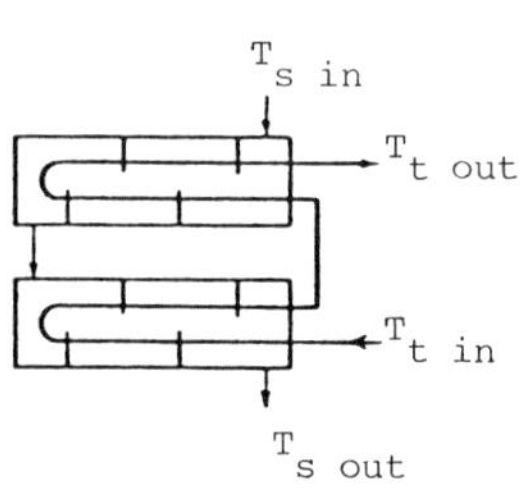

CORRECTION FACTOR TO COUNTERFLOW 'LMTD' FOR HEAT EXCHANGER WITH TWO SHELL PASSES AND A MULTIPLE OF TWO TUBE PASSES.

Fig. 8.10

8.13 HEAT EXCHANGER EFFECTIVENESS

Effectiveness or number of transfer units (NTU) method is used when the LMTD method gives rise to trial and error iteration.

$$\text{Effectiveness} = \varepsilon = \frac{\text{Actual heat transfer}}{\text{Maximum possible heat transfer}}$$

CASE I: For single pass, counter-flow exchanger

$$\varepsilon = \frac{1 - \exp\left[\frac{-UA}{C_{min}}\left(1 - \frac{C_{min}}{C_{max}}\right)\right]}{1 - \frac{C_{min}}{C_{max}} \exp\left[\frac{UA}{C_{min}}\left(1 - \frac{C_{min}}{C_{max}}\right)\right]} \qquad (8.7)$$

where

$$C_{min} = (mC_P)_c$$

$$C_{max} = (mC_P)_H$$

$$C_R = \frac{C_{min}}{C_{max}}$$

or 'ε' can be expressed in terms of NTU [number of transfer units]

$$NTU = \frac{UA}{C_{min}}$$

$$\therefore \quad \varepsilon = \frac{1 - \exp[-NTU(1 - C_R)]}{1 - C_R \exp[NTU(1 - C_R)]} \qquad (8.8)$$

CASE II: For single pass, parallel flow exchanger

$$\varepsilon = \frac{1 - \exp[-NTU(1 + C_R)]}{(1 + C_R)} \qquad (8.9)$$

Note: NTU depends on the size of the heat exchanger.

Nusselts number for a vertical tier:

$$Nu = \frac{hD}{k_i} = 0.725 \left(\frac{\rho_l(\rho_l - \rho_v)g(h_{fg})D^3}{Nu_l K_l \Delta T} \right)^{\frac{1}{4}} \qquad (8.10)$$

h_{fg} = Latent heat of condensation

ρ_l = Density of the liquids

ρ_v = Density of the vapors

D = Diameter (outside) of the tube

CHAPTER 9

TEMPERATURE DISTRIBUTION WITH MORE THAN ONE INDEPENDENT VARIABLE

9.1 UNSTEADY HEAT CONDUCTION IN SOLIDS

General Equation:

$$\rho C_P \cdot \frac{\partial T}{\partial t} = (\nabla \cdot K \nabla T) \tag{9.1}$$

If thermal conductivity is independent of the temperature or direction of flow, then

$$\frac{\partial T}{\partial t} = \alpha \cdot \nabla^2 T \tag{9.2}$$

where $\alpha = K/\rho C_P$ = Thermal diffusivity of solid.

9.2 HEATING OF A SEMI-INFINITE BODY OR SLAB

General Equation:

$$\frac{\partial \theta}{\partial t} = \alpha \cdot \frac{\partial^2 \theta}{\partial y^2}, \qquad \theta = \frac{T - T_o}{T_1 - T_o}$$

θ = Dimensionless quantity

Initial condition: at $t \leq 0$, $T = T_o$, $\theta = 0$ for all y.

Boundary Conditions:

(1) at $y = 0$ $\quad T = T_1 \quad \theta = 1 \quad$ for $t > 0$

(2) at $y = \alpha$ $\quad T = T_o \quad \theta = 0 \quad$ for $t > 0$

Hint: Heat conduction occurs only in y-direction.

Solution:

$$\boxed{\frac{T - T_o}{T_1 - T} = 1 - \mathrm{erf}\left(\frac{y}{\sqrt{4\alpha t}}\right)}$$

or

$$\frac{T - T_o}{T_1 - T} = \mathrm{erfC}\ \frac{y}{\sqrt{4\alpha t}} \tag{9.3}$$

9.3 HEATING OF A FINITE SLAB

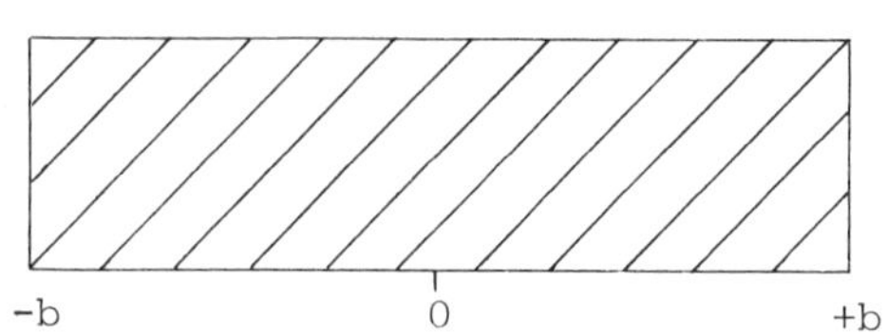

Fig. 9.1

At $t < 0 \quad T(\pm b, t) = T_o$

$t > 0 \quad T(\pm b, t) = T_1$

Introduce the dimensionless quantities

$$\theta = \frac{T_1 - T}{T_1 - T_o} = \text{dimensionless temperature}$$

$$\eta = \frac{y}{b} = \text{dimensionless length}$$

$$\tau = \frac{\alpha t}{b^2} = \text{dimensionless time}$$

General Equation:

$$\boxed{\frac{\partial \theta}{\partial \tau} = \frac{\partial^2 \theta}{\partial \eta^2}} \tag{9.4}$$

Initial Condition: at $\tau = 0 \quad \theta = 1$

Boundary Condition:

(1) at $\eta = +1 \quad \theta = 0$

(2) at $\eta = -1 \quad \theta = 0$

Solution:

$$\boxed{\frac{T_1 - T}{T_1 - T_o} = 2 \sum_{n=0}^{\infty} \frac{(-1)^n}{(n + \frac{1}{2})\pi} e^{-(n+\frac{1}{2})^2 \pi^2 \alpha t / b^2} \cos(n + \tfrac{1}{2}) \frac{\pi y}{b}} \tag{9.5}$$

Note that the solution of the problem is an infinite series. Such a series converges rapidly for large dimensionless time $\alpha t / b^2$.

9.4 COOLING A SPHERE IN CONTACT WITH WELL-STIRRED FLUID

A homogeneous solid sphere which is initially at a uniform temperature T_1 is suddenly immersed in a volume V_f of well-stirred fluid at a temperature T_o in an insulated tank. T_f denotes fluid temperature as a function of time. Assume perfect stirring.

Using dimensionless variables:

$$\theta_s = \frac{T_1 - T_s}{T_1 - T_o} = \text{dimensionless solid temperature.}$$

T_s is the temperature of solid as a function of time.

$$\theta_f = \frac{T_1 - T_f}{T_1 - T_o} = \text{dimensionless fluid temperature.}$$

$$\zeta = \frac{r}{R} = \text{dimensionless radial coordinate}$$

$$\tau = \frac{\alpha_s t}{R^2} = \text{dimensionless time}$$

$$\alpha_s = K_s / \rho_s C_{Ps} = \text{diffusivity of solid}$$

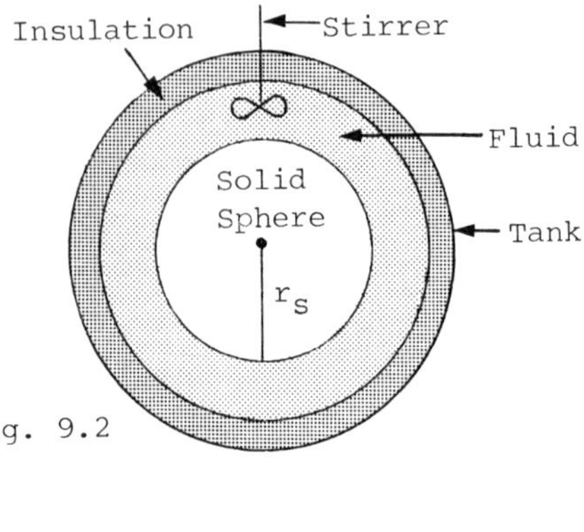

Fig. 9.2

9.4.1 GOVERNING DIFFERENTIAL EQUATIONS

SOLID REGION

General Equation:

$$\frac{\partial \theta}{\partial \tau} = \frac{1}{\zeta^2} \frac{\partial}{\partial \zeta} \left(\zeta^2 \frac{\partial \theta_s}{\partial \zeta} \right)$$

Boundary Conditions:

(1) at $\tau = 0$ $\quad \theta_s = 0$

(2) at $\zeta = 1$ $\quad \theta_s = \theta_f$

(3) at $\zeta = 0$ $\quad \theta_s = \text{finite}$

FLUID REGION

General Equation:

$$\frac{d\theta}{d\tau} = -\frac{3}{B} \left. \frac{\partial \theta_s}{\partial \zeta} \right|_{\zeta = 1}$$

Boundary Condition:

At $\tau = 0$ $\quad \theta_f = 1$

where

$$B = \frac{V_f \rho_f C_{Pf}}{V_s \rho_s C_{Ps}} = \text{dimensionless quantity}$$

V_s = volume of the fluid and of the solid.

9.4.2 SOLUTION OF REGIONAL EQUATIONS BY LAPLACE TRANSFORMATION

SOLID REGION

$$\boxed{P\overline{\theta}_s = \frac{1}{\zeta^2} \frac{d}{d\zeta} \left(\zeta^2 \frac{d\overline{\theta}_s}{d\zeta} \right)} \tag{9.6}$$

Boundary Condition:

(1) at $\zeta = 1$ $\qquad \bar{\theta}_s = \bar{\theta}_f$

(2) at $\zeta = 0$ $\qquad \bar{\theta}_s = \text{finite}$

FLUID REGION

$$P\bar{\theta}_f - 1 = -\frac{3}{B}\left.\frac{d\bar{\theta}}{d\zeta}\right|_{\zeta=1}$$

Solution:

$$\theta_f = \frac{B}{1+B} + 6B\sum_{K=1}^{\infty}\frac{e^{-b_K^2\tau}}{B^2 b_K^2 + 9(1+B)} \qquad (9.7)$$

where b_K are the non-zero roots of $\tan b = \dfrac{3b}{3 + Bb^2}$

9.5 THE BOUNDARY LAYER FLOW

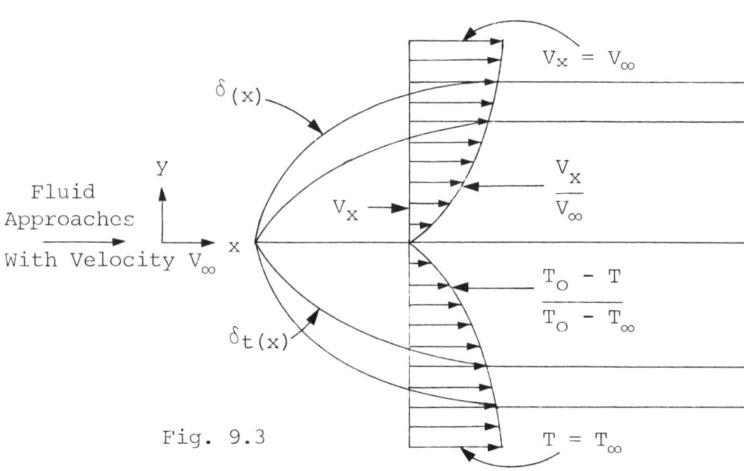

Fig. 9.3

$\delta(x)$ = Velocity boundary layer thickness

$V(x)$ = Velocity of free stream

$\delta_t(x)$ = Thermal boundary layer thickness

T_∞ = Fluid temperature far away from plate

t_s = Temperature of plate surface.

GOVERNING EQUATIONS FOR BOUNDARY LAYER FLOW

Restrictions:

(1) Flow is continuous

(2) Fluid should be Newtonian

(3) Steady state conditions

(4) Two-dimensional model

(5) No dissipation of heat

(6) Boundary layer flow: Streamlined bodies
Large Reynolds number

(7) Constant properties

(8) Flow in laminar region

(9) Applicable only within the boundary layer region.

Continuity Equation:

$$\frac{\partial v_x}{\partial x} + \frac{\partial v_y}{\partial y} = 0$$

X-Momentum:

$$v_x \frac{\partial v_x}{\partial x} + v_y \frac{\partial v_x}{\partial y} = -\frac{1}{\rho}\frac{dP}{dx}$$

$$= \nu \frac{\partial^2 v_x}{\partial y^2} + \beta g(t - t_f)$$

$$\frac{dP}{dy} = 0 \quad \text{or} \quad P = P(x)$$

Energy Equation:

$$v_x \frac{\partial t}{\partial x} + v_y \frac{\partial t}{\partial y} = \alpha \frac{\partial^2 t}{\partial y^2}$$

Momentum Integral Equation:

$$\frac{d}{dx} \int_0^{\delta} v\, dy - v_x \frac{d}{dx} \int_0^{\delta} v_x\, dy = - \frac{\delta}{\rho} \frac{dP}{dx} - \nu \left(\frac{\partial v_x}{\partial y}\right)_o \tag{9.8}$$

Energy Integral Equation:

$$\frac{d}{dx} \int_0^{\delta_t} (t_f - t) v_x\, dy = \alpha \left(\frac{\partial t}{\partial y}\right)_o \tag{9.9}$$

Solution of mementum and momentum integral equation are given in Chapter 5.

9.6 POHLHAUSEN'S SOLUTION FOR THE THERMAL

POLHAUSEN'S EQUATION

$$\phi'' + \frac{Pr}{2} \cdot f(\eta) - \phi' = 0$$

where
$$\phi = \frac{t - t_s}{t_f - t_s} = \phi(\eta)$$

Boundary Conditions:

(1) $\phi(0) = 0$

(2) $\phi(\infty) = 1$

Solution:

$$\phi(\eta) = \frac{t - t_s}{t_f - t_s} = \frac{\int_0^{\eta} (f'')^{Pr} d\eta}{\int_0^{\infty} (f'')^{Pr} d\eta} \tag{9.10}$$

Local heat transfer coefficient $h_x = K\,\phi'(0)\sqrt{\frac{V}{\nu x}}$

Average heat transfer coefficient $h = 2 \cdot \frac{K}{L}\,\phi'(0)\sqrt{\frac{VL}{\nu}}$

Local Nusselt number $NU_x = \phi'(0)\sqrt{\frac{Ux}{\nu}}$

Average Nusselt number $NU_L = 2 \cdot \phi'(0)\sqrt{\frac{UL}{\nu}}$

$$\phi'(0) \cong 0.332 Pr^{\frac{1}{3}} \quad \text{when } 0.6 < Pr < 15$$

Total heat transfer q_T

$$q_T = \int_0^L h(T_s - T_f) \cdot dA$$

$$= A(T_s - T_f)\,\frac{K}{L}\,NU_L$$

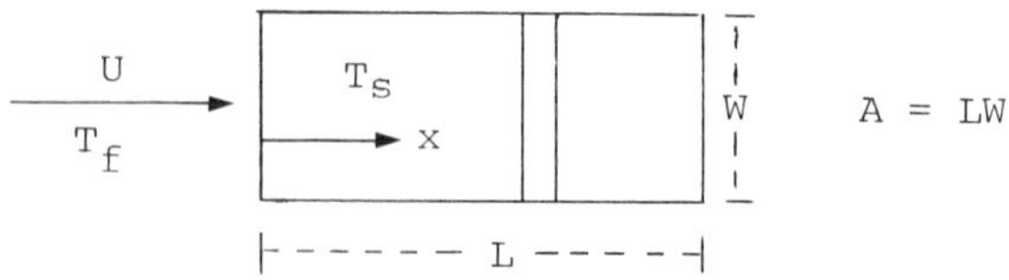

Fig. 9.4

9.7 TEMPERATURE PROFILE

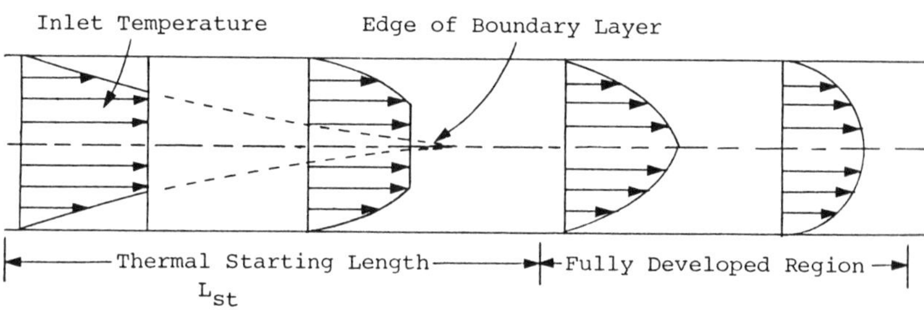

Development of Temperature Profile

Fig. 9.5

1. Bulk Fluid Temperature

Restrictions:

(a) Incompressible fluid

(b) Constant specific heat C_P

$$t_b = \frac{\int_0^R t v_r dr}{\int_0^R v_r \cdot r \cdot dr} \tag{9.11}$$

In many practical applications the following approximation is used

$$t_b = \frac{t_{inlet} + t_{outlet}}{2} \qquad (9.12)$$

2. The fully developed thermal profile is defined as

$$\frac{\partial}{\partial x}\left[\frac{t_s - t}{t_s - t_b}\right] = 0 \qquad (9.13)$$

3. Nusselt Number

(a) Constant-wall heat flux $(t_s - t_b)$ = constant

$$NU_D = \frac{h \cdot D}{K} = 4.36 \qquad (9.14)$$

(b) Constant wall temperature (t_s = constant)

$$NU_D = 3.66 \qquad (9.15)$$

4. Thermal starting length

$$\frac{L_{st}}{D} = 0.05R_{e\infty} Pr \qquad (9.16)$$

where L_{st} = Thermal starting length

D = Diameter of pipe

CHAPTER 10

FREE CONVECTION

10.1 CORRELATION DATA FOR VERTICAL PLANES AND CYLINDERS

(1) For Laminar Flow:

$$\frac{h_c L}{K_f} = 0.59 \left[\frac{L^3 \rho_f^2 g \beta_f \Delta t}{\mu_f^2} \left(\frac{C_p \mu}{K} \right)_f \right]^{0.25} \qquad (10.1)$$

where

$$\frac{L^3 \rho_f^2 g \beta_f \Delta t}{\mu_f^2} = N_{Grf} \Rightarrow \text{Grashoff Number}$$

$$\left(\frac{C_p \mu}{K} \right)_f = N_{Prf} \Rightarrow \text{Prandtl Number}$$

$$\frac{h_c L}{K_f} = N_{Nuf} \Rightarrow \text{Nusselt's Number}$$

Restriction:

$$10^4 < \frac{L^3 \rho g \beta_f \Delta t}{\mu_f^2} \left(\frac{C_p \mu}{K}\right)_f < 10^9$$

(2) For Turbulent Flow:

$$\boxed{N_{Nu} = 0.13(NG_{rf} \cdot NP_{rf})^{\frac{1}{3}}} \qquad (10.2)$$

Restriction:

$$10^9 < NG_{rf} \cdot NP_{rf} < 10^{12}$$

(3) For

$$NG_{rf} NP_{rf} < 10^4$$

use the logarithmic plot of N_{Nuf} vs. $NG_{rf} \cdot NP_{rf}$.

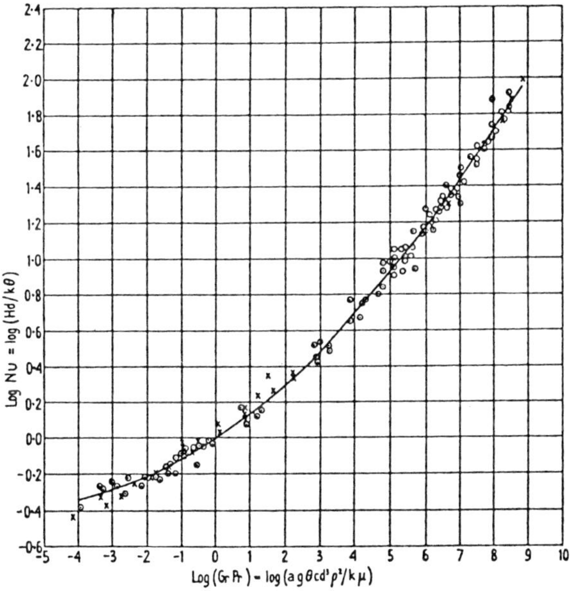

Natural convection for horizontal cylinders. ⊙s are experimental points for gases, Xs are experimental points for liquids.

Fig. 10.1

SIMPLIFIED EQUATION FOR AIR

For air at ordinary temperature at atmospheric pressure equation for vertical plane reduces to simplified dimensional equation.

(1) For $10^4 < NG_{rf} NP_{rf} < 10^9$

$$h_c = 0.19(\Delta t)^{\frac{1}{3}} \qquad (10.3)$$

(2) For $10^9 < NG_{rf} \cdot NP_{rf} < 10^{12}$

$$h_c = 0.29 \left(\frac{\Delta t}{L}\right)^{0.25} \qquad (10.4)$$

Putting $NG_{rf} \cdot NP_{rf} = x$

(1) $10^3 < x < 10^9$

$$h_c = 0.27 \left(\frac{\Delta t}{D_o}\right)^{0.25} \qquad (10.5)$$

(2) For $10^9 < x < 10^{12}$

$$h_c = 0.18\, \Delta t^{\frac{1}{3}} \qquad (10.6)$$

10.2 HORIZONTAL PLATES

(1) Heated plate facing upward or cooled plate facing downward in the laminar range.

$$\frac{hL}{K_f} = 0.54\left[\frac{L^3 \rho_f g \beta_f \Delta t}{\mu_f^2}\left(\frac{C_p \mu}{K}\right)_f\right]^{\frac{1}{3}} \qquad (10.7)$$

for $\quad 10^5 < NG_{rf} \cdot NP_{rf} < 2 \times 10^7$

(2) Turbulent Range

$$\frac{hL}{K_f} = 0.14\left[\frac{L^3 \rho_f g \beta_f \Delta t}{\mu_f^2}\left(\frac{C_p \mu}{K}\right)_f\right]^{\frac{1}{3}} \qquad (10.8)$$

for $\quad 2 \times 10^7 < NG_{rf} \cdot NP_{rf} < 3 \times 10^{10}$

(3) For heated plates facing downward, or cooled plates facing upward in laminar range.

$$\frac{h_c L}{K_f} = 0.27\left[\frac{L^3 \rho_f^2 g \beta_j \Delta t}{\mu_f}\left(\frac{C_p \mu}{K}\right)_f\right]^{0.25} \qquad (10.9)$$

for

$$3 \times 10^5 < \frac{L^3 \rho_f g \beta_j \Delta t}{\mu_f^2}\left(\frac{C_p \mu}{K}\right)_f < 3 \times 10^{10}$$

SIMPLIFIED EQUATION FOR AIR

(1) For heated plates facing up or cooled plates facing down in air at atmospheric pressure and at room temperature

a) Laminar range $h_c = 0.27\left(\frac{\Delta t}{L}\right)^{0.25}$

b) Turbulent Range $h_c = 0.22 \Delta t^{\frac{1}{3}}$

(2) For heated plates facing down, or cooled plates facing up

Laminar Range: $h_c = 0.12\left(\frac{\Delta t}{L}\right)^{0.25}$

10.3 FREE CONVECTION HEAT TRANSFER FROM A VERTICAL PLATE

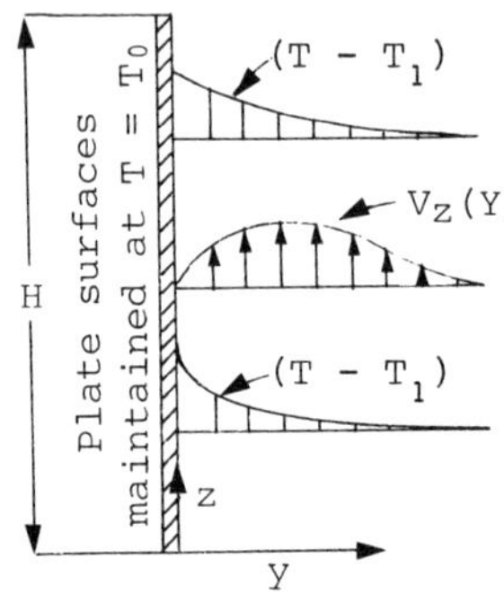

Temperature and velocity profiles in the neighborhood of a heated vertical plate.

Fig. 10.2

Assumption:

(1) $v_y \cong 0$

(2) $v_z = v_z(y,z)$

General Equation:

(1) Continuity: $$\frac{\partial v_y}{\partial y} + \frac{\partial v_z}{\partial z} = 0$$

(2) Motion: $$\rho\left(v_y \frac{\partial}{\partial z} + v_z \frac{\partial}{\partial z}\right) v_z = \mu\left(\frac{\partial^2 v_z}{\partial y^2} + \frac{\partial^2 v_z}{\partial z^2}\right) + \rho g P(T - T_1)$$

(3) Energy: $$\rho C_P\left(v_y \frac{\partial}{\partial y} + v_z \frac{\partial}{\partial z}\right)(T - T_1) = K\left(\frac{\partial^2}{\partial y^2} + \frac{\partial^2}{\partial z^2}\right)(T - T_1)$$

Boundary Conditions:

(1) At $y = 0$ $v_y = v_z = 0$ and $T = T_o$

(2) At $y = \infty$ $v_y = v_z = 0$ and $T = T_1$

(3) At $z = -\infty$ $T = T_1$ for all y

Introducing the dimensionless variables:

$$\nu = \frac{T - T_1}{T_o - T_1} \quad = \text{dimensionless temperature}$$

$$\zeta = \frac{Z}{H} \quad = \text{dimensionless vertical distance}$$

$$\eta = \left(\frac{B}{\mu \alpha H}\right)^{\frac{1}{4}} y \quad = \text{dimensionless horizontal distance}$$

$$\phi_z = \left(\frac{\mu}{B \alpha H}\right)^{\frac{1}{2}} v_z \quad = \text{dimensionless vertical velocity}$$

$$\phi_y = \left(\frac{\mu H}{\alpha^2 B}\right)^{\frac{1}{4}} v_y \quad = \text{dimensionless horizontal velocity}$$

where $\alpha = \dfrac{K}{\rho C_p}$ and $B = \rho g \beta (T_o - T_1)$

General equation reduces to:

Continuity: $$\frac{\partial \phi_y}{\partial \eta} + \frac{\partial \phi}{\partial \zeta} = 0$$

Motion: $$\frac{1}{Pr}\left(\phi_y \frac{\partial}{\partial \eta} + \phi_z \frac{\partial}{\partial \zeta} \right) \phi_z = \frac{\partial^2 \phi_z}{\partial \eta} + \nu$$

Energy: $$\phi_y \frac{\partial \nu}{\partial \eta} + \phi_z \frac{\partial \nu}{\partial \zeta} = \frac{\partial^2 \nu}{\partial \eta^2}$$

Boundary Conditions:

(1) At $\eta = 0$ $\phi_y = \phi_z = 0$ and $\nu = 1$

(2) At $\eta = \infty$ $\phi_y = \phi_z = 0$ and $\nu = 0$

(3) At $\zeta = -\infty$ $\nu = 0$ for all η

Solution:

$$q_{avg} = C \frac{K}{H} (T_o - T_1)(Gr \cdot Pr)^{\frac{1}{4}} \qquad (10.10)$$

$$C = \int_0^1 - \frac{\partial \nu}{\partial \eta}\bigg|_{\eta = 0} d\zeta$$

Function C is nearly a constant therefore it is termed as the weak function of the Prandtl number.

Pr	0.73 (air)	10	10^2	10^3
C	0.517	0.612	0.652	0.653

10.4 ENCLOSED SPACES

Heat transfer coefficient

$$h'_c = \frac{q_c}{A(T_{s_1} - T_{s_2})}$$

$$q_c = q - q_r$$

$q_c \Rightarrow$ heat transfer by convection

If natural convection were suppressed, the heat transferred only by conduction is given as

$$q_k = \frac{KA(t_{s1} - t_{s2})}{d} \qquad (10.11)$$

where d = clearance between the plates.

Then

$$\frac{q_c}{q_k} = \frac{h'_c A(t_{s1} - t_{s2})}{KA(t_{s_1} - t_{s_2})\Big|_d} = \frac{h'_c d}{K}$$

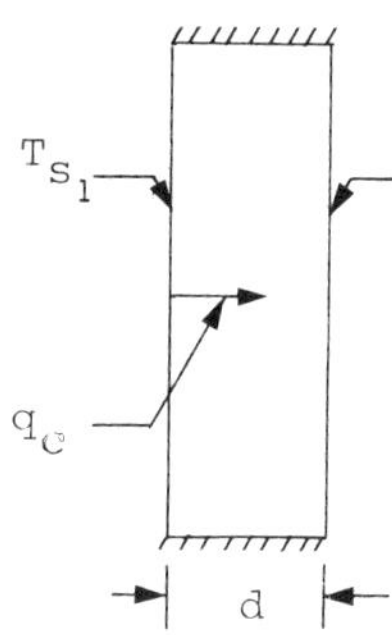

Fig. 10.3

VERTICAL ENCLOSED AIR SPACES

$$\frac{h'_c d}{K_f} = \frac{c}{\left(\frac{L}{D}\right)^{\frac{1}{9}}} \left[\frac{d^3 \rho_f^2 g \beta_f}{\mu_f^2} \left(\frac{C_p \mu}{K}\right)_f \right]^n \qquad (10.12)$$

$$\frac{d^3 \rho_f^2 g \beta_f}{\mu_f^2} = N_{Gr,d}$$

For $6 \times 10^3 < N_{Gr,d} \cdot Npr < 2 \times 10^5$

$$n = \frac{1}{4} \quad c = 0.2 \ \text{ for a given } \frac{L}{d}$$

For $2.1 \times 10^5 < N_{Gr,d} < 2 \times 10^7$

$$n = \frac{1}{3} \quad c = 0.073 \ \text{for a given } \frac{L}{d}$$

HORIZONTAL ENCLOSED AIR SPACE

$$\boxed{\frac{h_c d}{K_f} = C \left[\frac{d^3 \rho_f^2 g \beta_f \Delta t}{\mu_f^2} \left(\frac{C_p \mu}{K} \right)_f \right]^n} \qquad (10.13)$$

(1) For $1 \times 10^4 < N_{Gr,d} < 3 \times 10^5$ $\quad n = \frac{1}{4} \quad c = 0.21$

(2) For $3.2 \times 10^5 < N_{Gr,d} < 1 \times 10^7$ $\quad n = \frac{1}{3} \quad c = 0.075$

COEFFICIENT OF VOLUME EXPANSION 'β'

(1) $\beta = -\frac{1}{\rho}\left(\frac{\partial P}{\partial t}\right)_{P=\text{constant}} = \frac{1}{\nu}\left(\frac{\partial \nu}{\partial t}\right)_{P=\text{constant}}$

where $\nu = \frac{1}{\rho}$ = specific volume.

For an ideal gas

$$\rho = \frac{P}{RT} \quad \text{and} \quad \beta = \frac{1}{T}$$

where T is in absolute temperature.

T	English	SI	Conversion Factor
β	$\frac{1}{460 + °F} = \frac{1}{R}$	$\frac{1}{273.5 + °C} = \frac{1}{°K}$	$°K = \frac{5}{9} \times °R$ $°C = \frac{5}{9}(°F - 32)$

(2) Fluid velocity far away from the plate is zero. Near the plate there is some fluid motion due to the fact that the plate is at a temperature different from that of the fluid.

Driving Force: The density difference is the driving force that maintains the fluid motion and the convective heat transfer.

(3) Rayleigh Number 'Ra':

$$Ra = Gr \cdot Pr = \frac{l^3 g \beta \Delta t\, Pr}{v^2}$$

For the vertical flat plate problem, transition from laminar to turbulent flow occurs when the Rayleigh number equals approximately 10^9. This is then the critical Rayleigh number.

$$Ra_{L_c} = 10^9$$

where

$$L_c = \frac{10^9 v^2}{\beta g(t_s - t_f) \cdot Pr}$$

$x < L_c$ Laminar flow

$x > L_c$ Turbulent flow

10.5 LAMINAR FLOW: FREE CONVECTION BETWEEN TWO PARALLEL WALLS OF DIFFERENT TEMPERATURE

A fluid of density ρ and viscosity μ is placed between two vertical walls a distance 2b apart as shown in Figure 10.4. The heated wall at $y = -b$ is maintained at a temperature T_2 and the cooled wall at $y = +b$ is maintained at a temperature T_1. Because of the temperature gradient, the fluid near the hot wall rises and that near the cold wall descends.

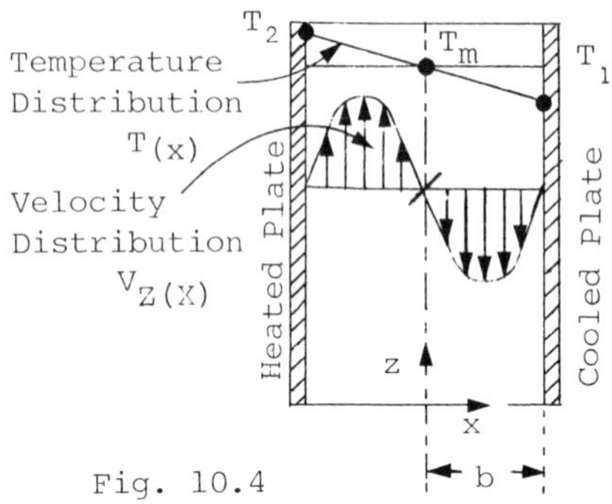

Fig. 10.4

Making a thermal energy balance over a shell of thickness Δx (for constant k), we obtain the following equation

$$K \frac{d^2T}{dx^2} = 0$$

Boundary Condition 1:

At $x = -b$ $\quad T = T_2$

At $x = +b$ $\quad T = T_1$

Solution:

$$\boxed{T = T_m - \frac{1}{2}\Delta T \left(\frac{x}{b}\right)} \qquad (10.14)$$

where $\Delta T = T_2 - T_1$

$$T_m = \frac{1}{2}(T_1 + T_2)$$

GENERAL EQUATION FOR VELOCITY DISTRIBUTION

$$\mu \frac{\partial^2 v_z}{\partial x^2} = \bar{\rho}\beta g \left[(T_m - T_R) - \frac{1}{2}\Delta T \left(\frac{x}{b}\right)\right]$$

where

T_R = Reference temperature

$\overline{\rho}$ = Density at temperature T_R

$\overline{\beta}$ = Coefficient of volume expansion evaluated at $T_R = \left.\frac{\partial P}{\partial T}\right|_{T_R}$

Boundary Condition:

(1) At $x = -b$ $\quad v_z = 0$

(2) At $x = +b$ $\quad v_z = 0$

Solution:

$$v_z = \frac{\rho\beta g b^2 \Delta T}{12\mu}(\eta^3 - \eta) \qquad (10.15)$$

where $\eta = \frac{y}{b}$

or

$$\phi = \frac{1}{12}\, Gr(\eta^3 - \eta)$$

where
$$\phi = \frac{b v_z \overline{\rho}}{\mu}$$

$$Gr = \left[\frac{\rho\beta\, g b^3 \Delta T}{\mu^2}\right]$$

10.6 CORRELATION FOR SPHERE

$$NU_D = 2 + 0.43\, Ra_\infty^{\frac{1}{4}} \qquad (10.16)$$

Restrictions:

(1) $Pr \approx 1$

(2) $1 \; R_{aD} < 10^5$

(3) Evaluate properties, except β at t_m

$$\beta \text{ at } \begin{cases} t_m & \text{for liquids} \\ t_m & \text{for gases} \end{cases}$$

Graetz Number:

$$\boxed{G_Z \equiv R_{eD} \cdot Pr \cdot \frac{D}{L}} \qquad (10.17)$$

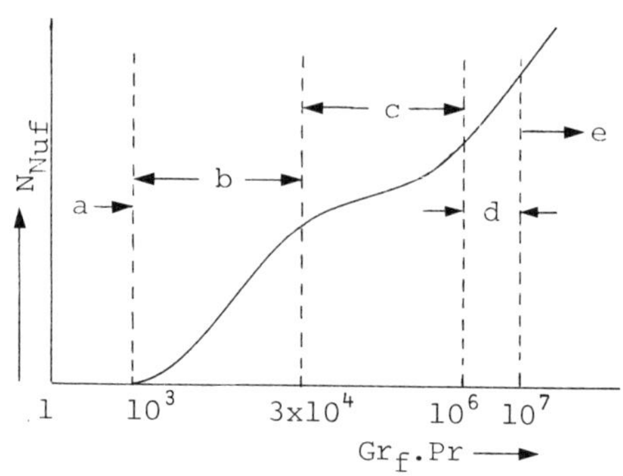

a → Conduction regime.
b → Asymptotic flow.
c → Laminar boundary layer flow.
d → Transition flow.
e → Turbulent boundary layer flow.

Fig. 10.5

CHAPTER 11

DIFFUSIVITY AND THE MECHANISM OF MASS TRANSFER

11.1 MOLECULAR DIFFUSION IN FLUIDS

This is the transfer of molecules (mass agregates) from one distinct phase to another or through a single phase due to a concentration gradient by means of the random motion of the molecules.

Ficks Law: The mass transport of an ith species per unit area is directly proportional to the concentration gradient in the direction of flow

$$j^*_i = \frac{m}{A} \alpha \nabla C_i$$

or

$$\boxed{j^*_{Ax} = -D_{AB} \frac{dC_A}{dx}} \qquad (11.1)$$

(For binary mixture of species A and B)

where

J^*_{Ax} = mass flux $\left(\frac{kg}{m^2\, hr}\right)$

$\frac{dC_A}{dx}$ = concentration gradient

D_{AB} = diffusion coefficient $\left(\frac{m^2}{hr}\right)$

11.1.1 MOLECULAR DIFFUSION IN GASES

$$-dP_A = \frac{RT}{D_{AB} \cdot P} (N_A P - N_A P_A - N_B P_A) dz \qquad (11.2)$$

where

$D_{AB} = \frac{R^2 T^2}{\beta P}$ = diffusivity of A to B

(1) Steady State Equimolal Counter Diffusion:

For equimolal counter diffusion N_A and N_B are constant.

$$\therefore \quad N_A = -N_B \quad \text{and} \quad Z = (Z_2 - Z_1)$$

$$\therefore \quad N_A = \frac{D_{AB}}{RTZ} (P_{A_1} - P_{A_2}) \qquad (11.3)$$

(2) Steady-State Diffusion of A Through Stagnant (non-diffusing) B:

$$N_B = 0 \qquad N_A = \text{constant}$$

$$N_A = \frac{D_{AB} P}{RTZP_{Bm}} = (P_{A_1} - P_{A_2}) \qquad (11.4)$$

where

$$P_{Bm} = \frac{P_{B_2} - P_{B_1}}{\ln\left(\frac{P_{B_2}}{P_{B_1}}\right)}$$

(3) Steady-State Diffusion of A Through a Stagnant Multicomponent Mixture:

$$N_A = \frac{D'_A P}{RTZ P_{BM}} (P_{A_1} - P_{A_2})$$

where

$$D'_A = \frac{1}{\frac{Y_B}{D_{AB}} + \frac{Y_C}{D_{AC}} + \frac{Y_D}{D_{AD}}}$$

Y = mole fraction component on an A-free basis.

11.1.2 DIFFUSIVITY OF GASES

$$D_{AB} = \frac{9.292 \times 10^{-4} T^{\frac{3}{2}} \frac{1}{M_A} + \frac{1}{M_B}}{P(\nu_{AB})^2 [f(KT/\varepsilon_{AB})} \tag{11.5}$$

D_{AB} = diffusivity, coefficient (cm²/sec)

T = absolute temperature (°K)

M_A, M_B = molar weight of A and B respectively

P = absolute pressure (atm)

ν_{AB} = molecular separation at collision [Å] $= \frac{\nu_A + \nu_B}{2}$

ε_{AB} = energy molecular interaction [ergs] $= \sqrt{\varepsilon_A \varepsilon_B}$

K = Boltzman constant = 1.38×10^{-16} erg/°K

$f(KT/\varepsilon_{AB})$ = a collision function given by Figure 11.1 or empirically

$$\frac{\varepsilon}{K} = 0.75\ T_c$$

$$\frac{\varepsilon}{K} = 1.39\ T_N$$

$$\nu = 0.835\ V_c^{\frac{1}{3}}$$

V_c = critical volume cm^3/g mol

T_c = critical temperature (°K)

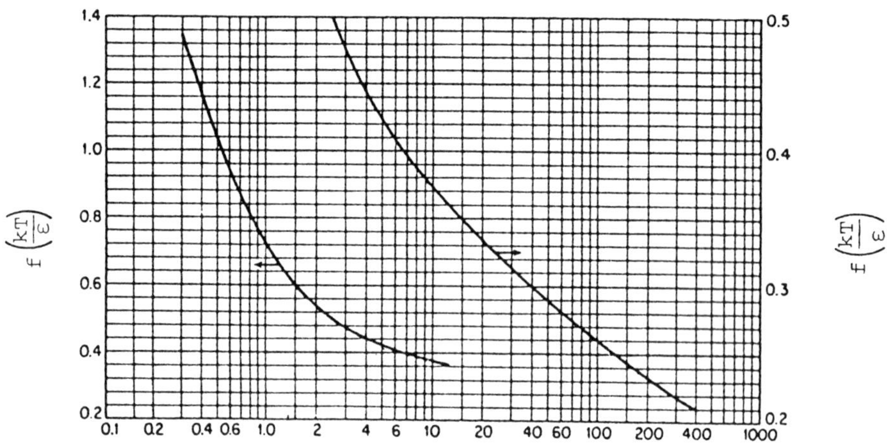

Fig.11.1

COLLISION FUNCTION FOR DIFFUSION

Force Constants of Gases:		
Gases	r, A	(ε /K)°K
Air	3.617	97.00
Helium	2.700	6.03
Neon	2.800	35.70
Argon	3.418	124.00
Nitrogen	3.681	91.46
Carbon-dioxide	3.996	190.00
Carbon monoxide	3.590	110.30
Nitric Oxide	3.470	119.00
Nitrous Oxide	3.879	220.30

11.1.3 MOLECULAR DIFFUSION IN LIQUIDS

$$j^*_{Ax} = -D_{AB} \frac{dC_A}{dx}$$

On integration, between two limits x_1 and x_2 with concentration C_{A_1} and C_{A_2} respectively.

$$j^* = -D_{AB} \frac{\Delta C_A}{\Delta x}$$

After integration

$$\boxed{j^*_{Ax} = D_{AB} \frac{(C_{A_1} - C_{A_2})}{x_2 - x_1}} \qquad (11.6)$$

(1) Steady-state Equimolal Counter Diffusion:

For equimolal counter diffusion

$$N_A = -N_B$$

$$\therefore \quad N_A = \frac{D_{AB}(C_{A_1} - C_{A_2})}{Z_2 - Z_1}$$

or

$$\boxed{N_A = \frac{D_{AB} C_{AV}(X_{A_1} - X_{A_2})}{Z_2 - Z_1}} \qquad (11.7)$$

C_{AV} = Average total concentration of A + B in kg mole/m^3

X_{A_1} and X_{A_2} are the mole fractions of A at 1 and 2 points.

(2) Diffusion of A Through a Stagnant B or Steady State:

N_A = constant $\quad N_B = 0$

$$D_{AB}(C_A + C_B) = D_{AB}C = \text{constant}$$

$$N_A = \frac{D_{AB}C(C_{A_1} - C_{A_2})}{C_{Bm} \cdot Z} = \frac{D_{AB}C(X_{A_1} - X_{A_2})}{X_{Bm} \cdot Z} \qquad (11.8)$$

C_{Bm} and X_{Bm} are the average concentration and mole fraction at position 1 and 2.

11.1.4 DIFFUSIVITY OF LIQUIDS

The quantity

$$D_{AB} = \frac{9.96 \times 10^{-16} T}{\mu v_A^{\frac{1}{3}}}$$

where T = Temperature in °K

μ = Viscosity of solution in kg/m • s

D_{AB} = diffusivity in m^2/S

v = molar volume = m^3/kg mole

The diffusivity can also be found for the case of diffusion of A in B, where A is the solute and B being the solvent.

$$D_{AB} = 1.173 \times 10^{-16} (\psi \mu_B)^{\frac{1}{2}} \frac{T}{\mu_B v_A^{0.6}} \qquad (11.9)$$

where ψ is an association parameter of the solvent.

Alternative method:

The quantity

$$\frac{T}{D_{AB} \mu_B} = F$$

where F is correlated as a function of the molal volume of the solute as shown in Figure 11.2.

$$\psi = \frac{\text{F for the diffusion in the solvent}}{\text{F for diffusion in water at constant solute molal volume}}$$

If the data for ψ is not available, ψ may be taken as 0.9.

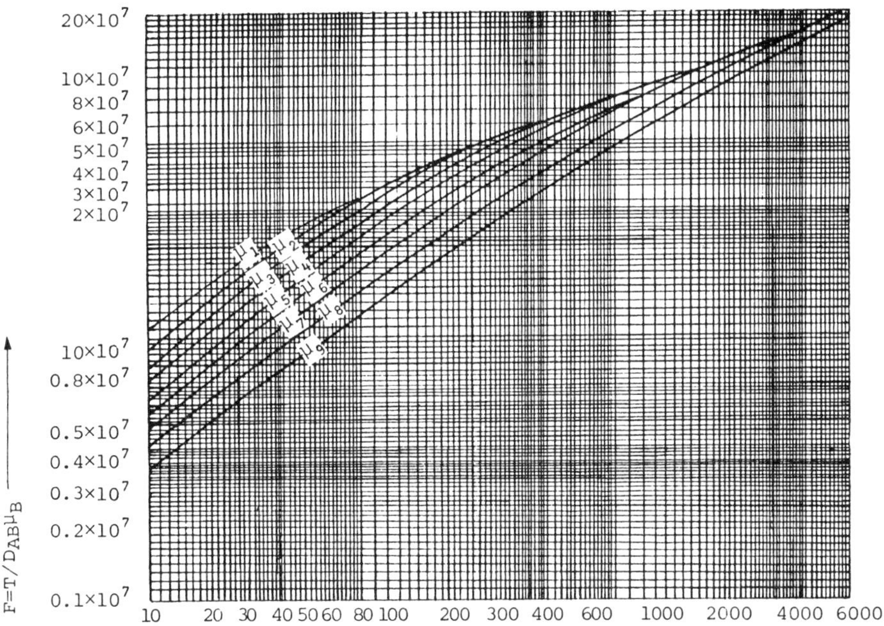

Fig. 11.2 V, molal volume of solute

11.1.5 DIFFUSION IN SOLIDS

Diffusion through a solid when the solute is dissolved to form a homogeneous solution may be termed structure insensitive diffusion.

The equation used for this type of diffusion is

$$N_A = -D_{AB} \frac{dC_A}{dZ}$$

If D_{AB} is constant, then on integration

$$N_A = \frac{D_{AB}(C_{A_1} - C_{A_2})}{Z} = K_s \Delta C_A \tag{11.10}$$

C_{A_1} and C_{A_2} are the concentration at opposite sides of the slab.

For solids with other shapes the rate is given by

$$\omega = N_A \cdot S_{Avg} = \frac{D_{AB} \, S_{Avg}(C_{A_1} - C_{A_2})}{Z} \tag{11.11}$$

where

S_{avg} = Average cross-section for diffusion.

For radial diffusion through a solid cylinder

$$S_{avg} = \frac{2\pi l(a_2 - a_1)}{\ln\left(\frac{a_2}{a_1}\right)}$$

where a_1 and a_2 are the inner and outer radii of the cylinder.

l = Length of the cylinder

$Z = a_2 - a_1$

For radial diffusion through a spherical shell

$$S_{avg} = 4\pi a_1 a_2$$

$$Z = a_2 - a_1$$

where a_1 and a_2 are the inner and outer radii.

11.2 UNSTEADY-STATE DIFFUSION

11.2.1 DIFFUSION FROM A SLAB WITH SEALED EDGES

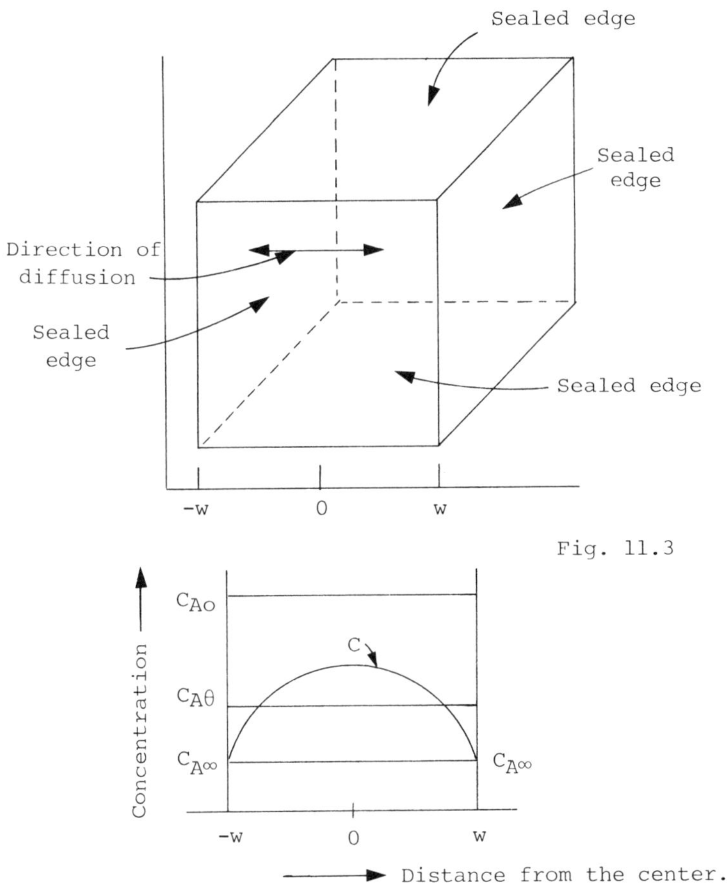

Fig. 11.3

Consider a slab of thickness 2W with four sealed edges, and the diffusion takes place only toward and from the flat parallel faces.

$C_{A_0} - C_{A_\infty}$ = Measure of the amount of solute removed

$C_{A\theta} - C_{A\infty}$ = Measure of the amount of solute still unremoved.

The fraction unremoved E is given by

$$E = \frac{C_{A\theta} - C_{A\infty}}{C_{A_0} - C_{A\infty}} = f\left(\frac{D\theta}{w^2}\right)$$

$$= \frac{8}{\pi}\left[e^{-D\theta\pi^2/4w^2} + \frac{1}{9}e^{-D\theta\pi^2 9/4w^2} + \frac{1}{25}e^{-D\theta 2\pi/4w^2} + \ldots +\right] = EW$$

11.2.2 DIFFUSION FROM A RECTANGULAR BAR WITH SEALED ENDS

$$E = f\left(\frac{D\theta}{w^2}\right) f\left(\frac{D\theta}{b^2}\right) = E_w E_b \qquad (11.12)$$

11.2.3 DIFFUSION IN A RECTANGULAR PARALLEL PIPE

For a bar of dimension 2w, 2b, 2L in which diffusion takes place from all six faces.

$$E = f\left(\frac{D\theta}{w^2}\right) f\left(\frac{D\theta}{b^2}\right) f\left(\frac{D\theta}{L^2}\right) = E_w E_b E_L \qquad (11.13)$$

11.2.4 DIFFUSION FROM A SPHERE

For a sphere of radius w

$$E = f'\left(\frac{D\theta}{w^2}\right) = E_Q \qquad (11.14)$$

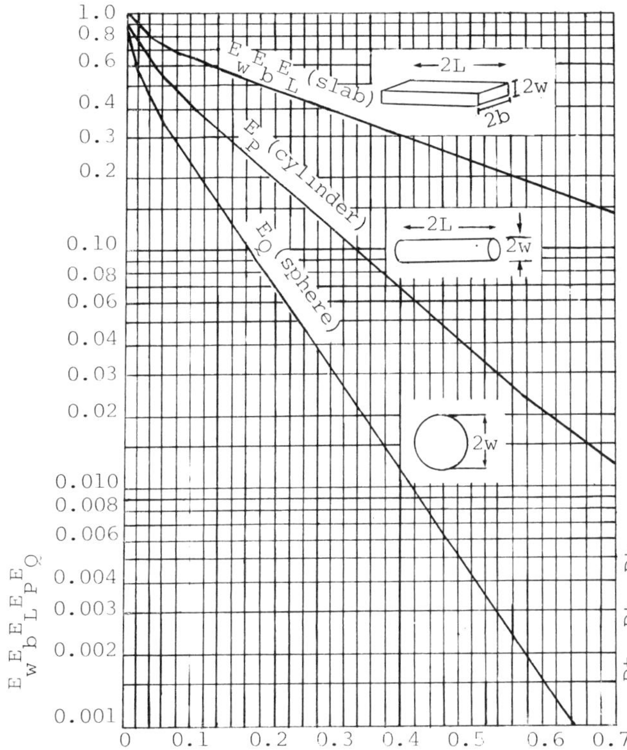

Fig. 11.4 Unsteady-state Diffusion

11.2.5 DIFFUSION FROM A CYLINDER WITH SEALED ENDS

For a cylinder of radius w with plane ends sealed

$$E = f''\left(\frac{D\theta}{w^2}\right) = E_P \qquad (11.15)$$

11.2.6 DIFFUSION FROM A CYLINDER

For a cylinder of radius w and length 2L with diffusion from both ends as well as from the cylindrical surface

$$\boxed{E = f\left(\frac{D}{c^2}\right) f''\left(\frac{D}{a^2}\right) = E_L E_P} \quad (11.16)$$

11.3 DEFINITION OF VELOCITIES, CONCENTRATIONS AND MASS FLUXES

(A) Mass Concentration $\rho_i = \frac{\text{Mass of species i}}{\text{Volume of solution}}$

(B) Molar Concentration $C_i = \frac{\text{No. of Moles of species i}}{\text{Volume of solution}} = \frac{\rho_i}{M_i}$

where M_i = molar weight of species i.

(C) Mass fraction 'ω_i' $\omega_i = \frac{\text{Mass concentration of species i}}{\text{Total mass density of solution}}$

$$= \frac{\rho_i}{\rho}$$

(D) Mole fraction 'x_i'

$$x_i = \frac{\text{Molar concentration of species i}}{\text{Total molar density of solution}} = \frac{C_i}{C}$$

(E) Local mass average velocity $v = \frac{\sum_{i=1}^{n} \rho_i v_i}{\sum_{i=1}^{n} \rho_i}$

where v_i = velocity of the ith species with respect to stationary coordinate axes.

(F) Local molar average velocity = $V^* = \dfrac{\sum_{i=1}^{n} C_i v_i}{\sum_{i=1}^{n} C_i}$

(G) Mass flux relative to stationary coordinates axes

$$n_i = \rho_i v_i$$

(H) Molar flux relative to stationary coordinate axes

$$N_i = c_i v_i$$

(I) Mass flux relative to the mass average velocity v

$$j_i = \rho_i (v_i - v)$$

(J) Molar flux relative to the mass average velocity v

$$J_i = c_i (v_i - v)$$

(K) Mass flux relative to the molar average velocity V*

$$j^* = \rho_i (v_i - v^*)$$

(L) Molar flux relative to the molar average velocity v*

$$J_i^* = c_i (v_i - V^*)$$

11.4 EQUIVALENT FORMS OF FICK'S LAW OF DIFFUSION

(1) $n_A - \omega_A(n_A + n_B) = -\rho D_{AB} \nabla\omega_A$

(2) $N_A - X_A(N_A + N_B) = -C D_{AB} \nabla X_A$

(3) $j_A = -\rho D_{AB} \nabla\omega_A$

(4) $J^*_A = -c D_{AB} \nabla X_A$

(5) $j_A = -\left(\frac{C^2}{\rho}\right) M_A \cdot M_B D_{AB} \nabla X_A$

(6) $J^*_A = -\left(\frac{\rho^2}{CM_AM_B}\right) D_{AB} \cdot \nabla\omega_A$

(7) $C(V_A - V_B) = -\frac{CD_{AB}}{X_AX_B}(\nabla X_A)$

where $\nabla\omega_A = \frac{d\omega_A}{dZ}$, ω_A = weight fraction of A

$$\nabla X_A = \frac{dX_A}{dZ}$$

V_A = velocity of A relative to stationary coordinates m/s.

N_A = molar flux of A

N_B = molar flux of B.

CHAPTER 12

CONVECTIVE MASS TRANSFER

12.1 MASS TRANSFER COEFFICIENTS

Definition:

$$J^*_{A_1} = K'_C (C_{A_1} - C_{A_2})$$

$J^*_{A_1}$ = flux of A from the surface A_1 relative to the whole bulk phase

K'_C = convective mass-transfer coefficient

$$K'_C = \frac{D_{AB} + \bar{\varepsilon}_M}{Z_2 - Z_1}$$ = experimental mass transfer coefficient

where

$\bar{\varepsilon}_M$ = Mass eddy diffusion in m^2/s.

C_{A_1} and C_{A_2} are the concentration of points (1) and (2).

12.1.1 MASS TRANSFER COEFFICIENT FOR EQUIMOLAR COUNTER DIFFUSION

$$N_A = -N_B \qquad N_A = K_{C'} (C_{A_1} - C_{A_2})$$

where
$$K_{C'} = \frac{D_{AB} + \overline{\varepsilon}_M}{Z_2 - Z_1}$$

Let Y_A = mole fraction in gas phase

X_A = mole fraction in liquid phase

Then for equimolar-counter diffusion:

Gases:
$$N_A = K_C'(C_{A_1} - C_{A_2}) = K_G'(P_{A_1} - P_{A_2})$$
$$= K_Y(Y_{A_1} - Y_{A_2})$$

Liquids:
$$N_A = K_C'(C_{A_1} - C_{A_2}) = K_L'(C_{A_1} - C_{A_2})$$
$$= K_X(X_{A_1} - X_{A_2})$$

Relationship between the Mass-transfer Coefficients:

Gases:
$$K_C'C = K_C'\frac{P}{RT} = K'y = K_C\frac{P_BM}{RT}$$

Liquids:
$$K_C'C = K_L'C = K_L\ \rho/M = K_X'$$

ρ = Density of the liquids

M = Molecular weight of the liquid.

12.1.2 MASS TRANSFER COEFFICIENT FOR A THROUGH STAGNANT B

Here $N_B = 0$

$$\therefore \quad N_A = K_C(C_{A_1} - C_{A_2}) = \frac{K_C'}{X_BM}(C_{A_1} - C_{A_2})$$

$$K_C = \frac{K_C'}{X_BM} = K_C'\ \frac{\ln \dfrac{X_{B_2}}{X_{B_1}}}{X_{B_2} - X_{B_1}}$$

Gases: $N_A = K_C(C_{A_1} - C_{A_2}) = K_G(P_{A_1} - P_{A_2})$

$= K_Y(Y_{A_1} - Y_{A_2})$

Liquids: $N_A = K_C(C_{A_1} - C_{A_2}) = K_L(C_{A_1} - C_{A_2})$

$= K_X(X_{A_1} - X_{A_2})$

Conversion between mass-transfer coefficients:

Gases: $K_C \frac{P_{BM}}{RT} = K_G P_{BM} = K_Y \cdot Y_{BM}$

$= K_C Y_{BM} \cdot C$

Liquids: $K_L X_{BM} C = K_X \cdot X \cdot B_M$

12.2 MASS TRANSFER FOR FLOW INSIDE PIPES

12.2.1 MASS TRANSFER FOR LAMINAR FLOW INSIDE PIPES

For Gases:

$$N_{Re} < 2100 \qquad \text{(Laminar flow)}$$

Data obtained from mass transfer from the walls for gases are plotted in the Figure 12.1 for

$$\left(\frac{w}{D_{AB} \cdot \rho \cdot L}\right) < 70$$

$$\boxed{\frac{w}{D_{AB}\rho L} = N_{Re} \cdot N_{Sc}\left(\frac{D}{L}\right)\left(\frac{\pi}{4}\right)} \qquad (12.1)$$

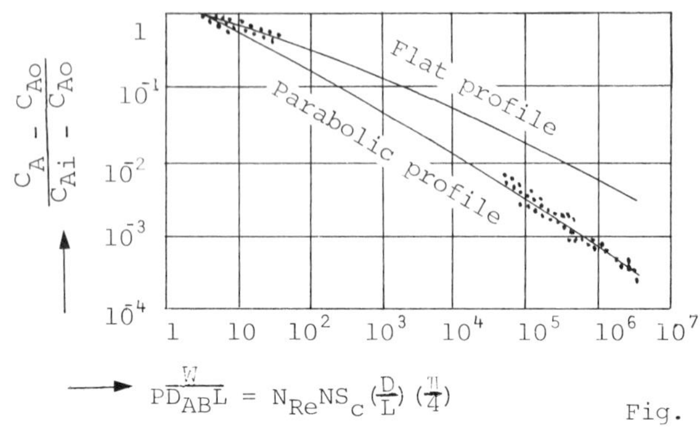

Fig. 12.1

C_A = Exit concentration

C_{Ai} = Concentration at the interface

C_{Ao} = Inlet concentration

For Liquids:

$$\frac{w}{D_{AB}\,\rho L} > 400$$

Data follow the parabolic line, for small values of D_{AB}

$$\frac{C_A - C_{Ao}}{C_{Ai} - C_{Ao}} = 5.5\left(\frac{w}{D_{AB}\,\rho L}\right)^{-\frac{2}{3}} \qquad (12.2)$$

12.2.2 MASS TRANSFER FOR TURBULENT FLOW INSIDE PIPES

$N_{Re} > 2100$ (Turbulent flow)

$$N_{Sh} = 0.023\left(\frac{Dv\rho}{\mu}\right)^{0.83}\left(\frac{\mu}{\rho D_{AB}}\right)^{0.33} \qquad (12.3)$$

Restrictions:

(1) $N_{Re} > 2100$

(2) $0.6 \leq N_{Sh} \leq 3000$

12.2.3 MASS TRANSFER FOR FLOW INSIDE WETTED WALL TOWERS

(A) For Gases:

Laminar flow:

$$\frac{C_A - C_{Ao}}{C_{Ai} - C_{Ao}} = 5.5 \left(\frac{w}{D_{AB} \rho L} \right)^{-\frac{2}{3}} \qquad (12.4)$$

Restrictions:

(1) $N_{Re} < 2100$

(2) $\frac{w}{D_{AB} \rho L} < 70$

Turbulent flow:

$$N_{Sh} = 0.023 \left(\frac{D v \rho}{\mu} \right)^{0.83} \left(\frac{\mu}{\rho D_{AB}} \right)^{0.33} \qquad (12.5)$$

Restrictions:

(1) $N_{Re} > 2100$

(2) $0.6 \leq N_{Sh} \leq 3000$

(B) For Liquids:

Laminar flow:

$$N_{AX}(Z)\Big|_{X=0} = C_{Ao}\sqrt{\frac{D_{AB}V_{max}}{\pi Z}} \quad (12.6)$$

Restriction:

$$N_{Re} = \frac{4\rho\delta V_{Z,\,ave}}{\mu} \leq 1200$$

For turbulent flow:

$$N_A = L\,C_{Ao}\sqrt{\frac{4D_{AB}V_{max}}{\pi L}} \quad (12.7)$$

12.3 MASS TRANSFER FOR FLOW OUTSIDE SOLID SURFACE

12.3.1 MASS TRANSFER IN FLOW PARALLEL TO FLAT PLATES

For gases:

(A) $J_D = 0.664(N_{Re}L)^{-0.5}$

or

$$\frac{K_C'L}{D_{AB}} = N_{Sh} = 0.664(N_{Re,L})^{0.5}(N_{Sh})^{\frac{1}{3}} \quad (12.8)$$

Restrictions:

(1) $N < 1500$

(2) $J_D = J_H = f/2$

(B) $J_D = 0.036(N_{Re,L})^{-0.2}$

Restriction:

$$15{,}000 \leq N_{Re,L} \leq 300{,}000$$

For Liquids:

$$J_D^* = 0.99\, N_{Re,L}^{-0.5} \qquad (12.9)$$

Restriction:

$600 \leq N_{Re,L} \leq 50{,}000$

12.3.2 MASS TRANSFER FOR FLOW PAST A SINGLE SPHERE

For gases:

$$N_{Sh} = 2 + 0.552\, N_{Re}^{0.53} N_{Sc}^{\frac{1}{3}} \qquad (12.10)$$

Restrictions:

(1) $0.6 \leq N_{Sh} \leq 2.7$

(2) $1 \leq N_{Re} \leq 48{,}000$

For liquids:

$$N_{Sh} = 2 + 0.95\, N_{Re}^{0.5} \; N_{Sc}^{\frac{1}{3}} \qquad (12.11)$$

Restriction:

$$2 \leq N_{Re} \leq 2{,}000$$

$$\boxed{N_{Sh} = 0.347\, N_{Re}^{0.62}\, N_{Sc}^{\frac{1}{3}}} \qquad (12.12)$$

Restrictions:

$$2{,}000 \leq N_{Re} \leq 17{,}000$$

$$N_A = \frac{2D_{AB}}{D_P}(C_{A1} - C_{A2}) = K_C(C_{A1} - C_{A2})$$

Mass transfer coefficient K_C which K'_C for a dilute solution is then

$$K'_C = 2\,\frac{D_{AB}}{D_P}$$

$$\frac{K'_C D_P}{D_{AB}} = N_{Sh} = 2.0$$

where D_P is the diameter of the sphere.

12.3.3 MASS TRANSFER IN PACKED BEDS

For Gases:

$$\boxed{J_D = J_H = \frac{0.4548}{\varepsilon}\, N_{Re}^{-0.4069}} \qquad (12.13)$$

$$N = \frac{D_P v' \rho}{\mu}$$

D_P = Diameter of the sphere

v' = Superficial mass average velocity in the empty tube that is without packing

ε = Void fraction in a bed (m^3)

Restrictions:

(1) $10 \leq N_{Re} \leq 10{,}000$

(2) $0.3 \leq \varepsilon \leq 0.5$

For Liquids:

(A) $$J_D = \frac{1.09}{\varepsilon} N_{Re}^{-\frac{2}{3}} \qquad (12.14)$$

Restrictions:

(1) $0.0016 \leq N_{Re} \leq 55$

(2) $165 \leq N_{Sc} \leq 70{,}600$

(B) $$J_D = \frac{0.250}{\varepsilon} N_{Re}^{-0.31} \qquad (12.15)$$

Restrictions:

(1) $55 \leq N_{Re} \leq 1500$

(2) $165 \leq N_{Sc} \leq 10{,}690$

12.3.4 FLUIDIZED BED

For Liquids:

$$\varepsilon J_D = 1.1068 N_{Re}^{-0.72} \qquad (12.16)$$

Restriction:

$$1 \leq N_{Re} \leq 10$$

For Gases and Liquids:

$$\boxed{J_D = J_H = \frac{0.4548}{\varepsilon} N_{Re}^{-0.4069}} \tag{12.17}$$

Restriction:

$$10 \leq N_{Re} \leq 41,000$$

12.3.5 MASS TRANSFER FOR FLOW PAST SINGLE CYLINDER

$$\boxed{J_D = 0.600(N)^{-0.487}} \tag{12.18}$$

Restrictions:

(1) $0.6 \leq N \leq 2.6$ for gases

(2) $1000 \leq N \leq 3000$ for liquids

(3) $50 \leq N \leq 50,000$

(4) Cylinder should be long

(5) Mass transfer at the ends of the cylinder is negligible.

Cylinders:

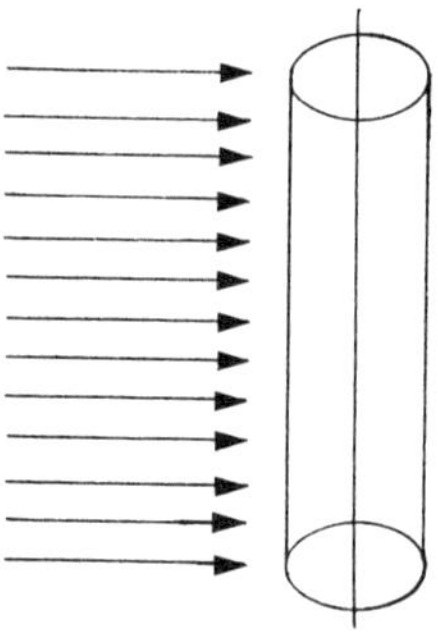

Fig. 12.2

Flow Perpendicular to Cylinder

12.4 MASS TRANSFER TO SUSPENSIONS OF SMALL PARTICLES

12.4.1 MASS TRANSFER TO SMALL PARTICLES LESS THAN 0.6mm

$$K'_L = \frac{2D_{AB}}{D_P} + 0.37\, N_{Sc}^{-\frac{2}{3}} \left(\frac{\Delta \rho \mu_c g}{\rho_c^2} \right)^{\frac{1}{3}} \qquad (12.19)$$

D_{AB} = Diffusivity of solute A in solution B

D_P = Diameter of gas bubble or the solid particle

μ_C = Viscosity of the solution

$\Delta\rho$ = $(\rho_i - \rho_P)$ or $(\rho_P - \rho_C)$ the value of $\Delta\rho$ is always positive

ρ_C = Density of the continuous phase

ρ_P = Density of gas or solid particle.

12.4.2 MASS TRANSFER TO LARGE GAS BUBBLE GREATER THAN 2.5mm

$$K'_L = 0.42\, N_{Sc} \left(\frac{\Delta \rho \mu_C g}{\rho_c^2} \right)^{\frac{1}{3}} \qquad (12.20)$$

This usually takes place in mixing vessels, and K'_L is independent of bubble size.

12.4.3 MASS TRANSFER TO PARTICLES IN HIGHLY TURBULENT MIXERS

$$K'_L = N_{Sc}^{\frac{2}{3}} = 0.13 \left(\frac{(P/v)\,\mu_C}{\rho_C^2} \right)^{\frac{1}{4}} \qquad (12.21)$$

P/v = Power input per unit volume

Here the mass transfer takes place due to the force of gravity caused by the density difference between phases.

CHAPTER 13

CONCENTRATION DISTRIBUTION IN SOLIDS AND IN LAMINAR FLOW

13.1 SHELL MASS BALANCE

Mass transfer and diffusion problems can be solved by making a shell mass balance. The procedure is as follows:

(1) Consider a thin shell of thickness (Δx) perpendicular to the direction of mass transfer. Make a mass balance over this shell and obtain a first order differential equation. Solution of this equation gives the mass-flux distribution.

(2) Substitute the relation for Fick's Law into this differential equation and obtain a second order differential equation. Solution of this second order equation gives the concentration distribution. The constant appearing in the integrated equation can be obtained by using boundary conditions.

(3) The Law of Conservation of Mass gives mass balance on A as,

(Rate of mass A in) - (Rate of mass A out)

+ (Rate of generation of A)

= (Rate of accumulation of mass A)

Boundary conditions:

(a) At the surface, the concentration

$$C_A = C_{Ao} \quad \text{or} \quad X_A = X_{Ao}$$

(b) Mass flux at the boundary

$$N_A = N_{Ao}$$

Note that moles can also be considered instead of mass.

(c) When diffusion is occuring in a solid substance A may be lost to a surrounding fluid according to the relation

$$N_{Ao} = K_C(C_{Ao} - C_{Af})$$

where

N_{Ao} = Mass flux at the surface

C_{Ao} = Surface concentration

C_{Af} = Fluid stream concentration

K_C = Mass transfer coefficient.

(d) The rate of dissappearance of A can be specified. For example

$$N_{Ao} = K''C_A$$

K'' = First order chemical rate constant.

13.2 DIFFUSION THROUGH A STAGNANT GAS FILM

Assumptions:

(1) T = constant temperature

(2) P = total constant pressure

(3) Steady state diffusion

(4) Mole fraction X_{P_2} = constant

(5) Solubility of Q in liquid P is negligible

(6) Both gases P and Q behave as an ideal gas

(7) $X_{P_1} + X_{Q_1} = 1$ or $X_{P_2} + X_{Q_2} = 1$

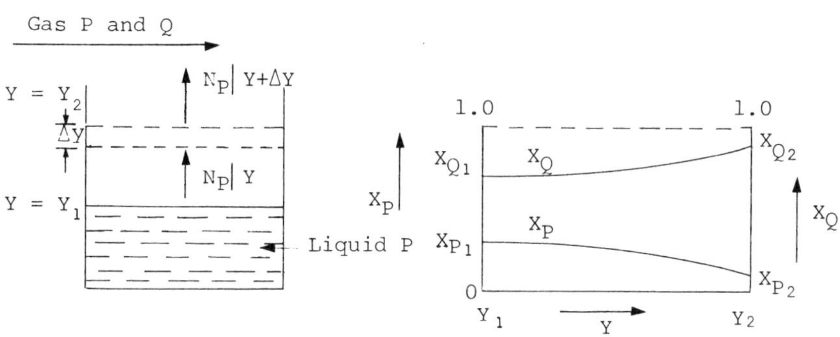

Fig. 13.1

General Equation:

$$-\frac{dN_{PY}}{dY} = 0$$

where

$$N_{PY} = -c\,\frac{D_{PQ}}{1 - X_P}\,\frac{dx_P}{dY}$$

Solution:

$$\left(\frac{1 - X_P}{1 - X_{P_1}}\right) = \left(\frac{1 - X_{P_2}}{1 - X_{P_1}}\right)^{\frac{Y - Y_1}{Y_2 - Y_1}} \tag{13.1}$$

or

$$\frac{X_Q}{X_{Q_1}} = \left(\frac{X_{Q_2}}{X_{Q_1}}\right)^{\frac{Y - Y_1}{Y_2 - Y_1}}$$

The average concentration of Q is given by

$$\frac{X_{Q,\,avg}}{X_{Q_1}} = \frac{\int_{Y_1}^{Y_2} \left(\frac{X_Q}{X_{Q_1}}\right) dy}{\int_{Y_1}^{Y_2} dy}$$

$$X_{Q,\,avg} = \frac{X_{Q_2} - X_{Q_1}}{\ln\left(\frac{X_{Q_2}}{X_{Q_1}}\right)} \tag{13.2}$$

The mass transfer rate:

$$N_{PY}\Big|_{Y=Y_1} = \frac{CD_{PQ}}{(Y_2 - Y_1)(X_Q)_{ln}} (X_{P_1} - X_{P_2}) \tag{13.3}$$

where

$$(X_Q)_{ln} = \frac{X_{Q_2} - X_{Q_1}}{\ln \frac{X_{Q_2}}{X_{Q_1}}} = \frac{X_{P_1} - X_{P_2}}{\ln \frac{X_{Q_2}}{X_{Q_1}}}$$

In terms of partial pressure and total pressure.

$$N_{PY}\Big|_{Y=Y_1} = \frac{PD_{PQ}/RT}{Y - Y(PQ)_{ln}} (P_{P_1} - P_{P_2}) \quad (13.4)$$

13.3 DIFFUSION WITH HETERO GENEOUS CHEMICAL REACTION

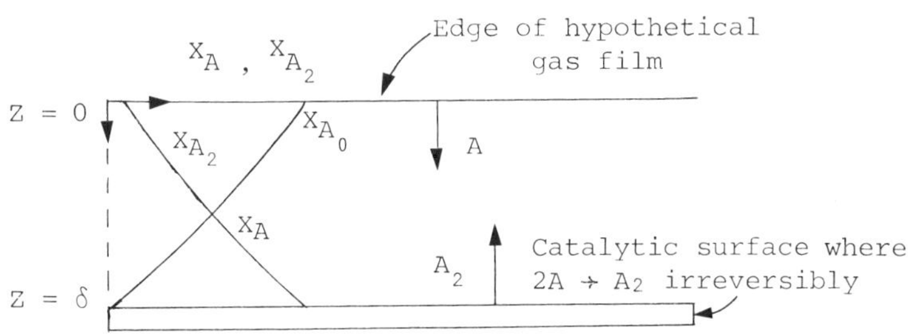

IDEALIZED PICTURE OF THE DIFFUSION PROBLEM NEAR A CATALYST PARTICLE.

Fig. 13.2

Assumptions:

(1) There is a gas film which surrounds each catalyst particle and A has to diffuse into the gas film to reach the catalyst surface.

(2) $2A \rightarrow A_2$ is an instantaneous and irreversible reaction at the catalyst surface.

(3) The gas film has a thickness of "δ".

(4) The temperature of the gas film is constant.

(5) The above reaction is diffusion controlled.

General Equation:

$$\frac{dN_{AZ}}{dZ} = 0$$

$$N_{AZ} = -\frac{1}{2} N_{AZ}$$

Boundary Condition:

(1) at $Z = 0$ $\quad X_A = X_{Ao}$

(2) at $Z = \delta$ $\quad X_A = 0$

Solution:

(1) Concentration profile:

$$\left(1 - \frac{1}{2} X_A\right) = \left(1 - \frac{1}{2} X_{Ao}\right)^{1 - \frac{Z}{\delta}} \qquad (13.5)$$

(2) Molar flux through the film:

$$\boxed{N_{AZ} = \frac{2CD_{AA_2}}{\delta} \ln \left(\frac{1}{1 - \frac{1}{2} X_{Ao}}\right)} \qquad (13.6)$$

When the reaction is not instantaneous:

Assumption:

$$N_{AZ} = K_1''C_A = CK_1''X_A$$

where $C_A = CX_A$

General Equation:

$$\frac{dN_{AZ}}{dZ} = 0 \qquad N_{A_2Z} = -\frac{1}{2} N_{AZ}$$

Boundary Conditions:

(1) at $Z = 0$ $X_A = X_{Ao}$

(2) at $Z = \delta$ $X = N_{AZ}/CK_1''$

Solution:

(1) Concentration profile:

$$\left(1 - \frac{1}{2} X_A\right) = \left(1 - \frac{1}{2} \frac{N_{AZ}}{CK_1''}\right)^{\frac{Z}{\delta}} \left(1 - \frac{1}{2} X_{Ao}\right)^{1-\frac{Z}{\delta}}$$

(2)

$$N_{AZ} = \frac{2C D_{AA_2}}{\delta} \ln \left[\frac{1 - \frac{1}{2} \frac{N_{AZ}}{CK_1''}}{1 - \frac{1}{2} X_{Ao}}\right] \quad (13.7)$$

13.4 DIFFUSION WITH HOMOGENEOUS CHEMICAL REACTION

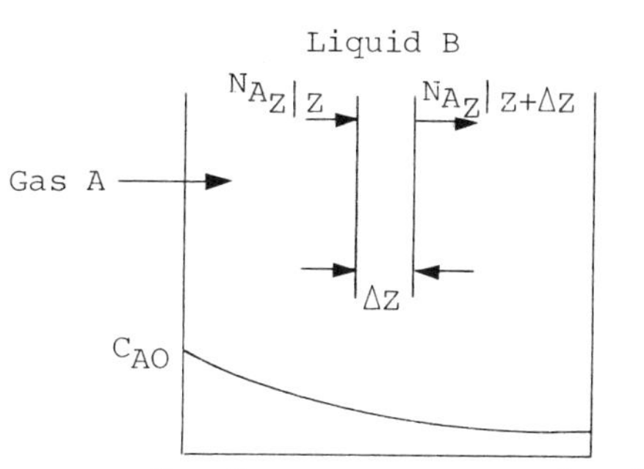

Fig. 13.3

Consider a system where gas A diffuses into the liquid phase B.

The reaction is considered to be first order and irreversible

$$A + B \rightarrow AB$$

General Equation:

$$-D_{AB} \frac{d^2C_A}{dZ^2} + K_1''' C_A = 0$$

K_1''' = First order rate constant for chemical reaction

Boundary Conditions:

(1) at $Z = 0$ $\quad C_A = C_{Ao}$

(2) at $Z = L$ $\quad N_{AZ} = 0$ or $\frac{dC_A}{dZ} = 0$

Solution:

(1) Concentration profile is given by

$$\boxed{\frac{C_A}{C_{Ao}} = \frac{\cosh b_1[(1 - \frac{Z}{L})]}{\cosh b_1}} \tag{13.8}$$

where

$$b_1 = \sqrt{\frac{K_1''' L^2}{D_{AB}}}$$

(2) Average concentration of A in liquid phase is given by

$$\boxed{\frac{C_{A,avg}}{C_{Ao}} = \frac{\int_0^L (C_A/C_{Ao})dZ}{\int_0^L dZ} = \frac{1}{b_1} \tanh b_1} \tag{13.9}$$

(3) The molar flux of A at plane Z = 0

$$N_{AZ}\Big|_{Z=0} = \left(\frac{D_{AB}C_{Ao}}{L}\right) b_1 \tanh b_1 \qquad (13.10)$$

13.5 DIFFUSION INTO A FALLING LIQUID FILM

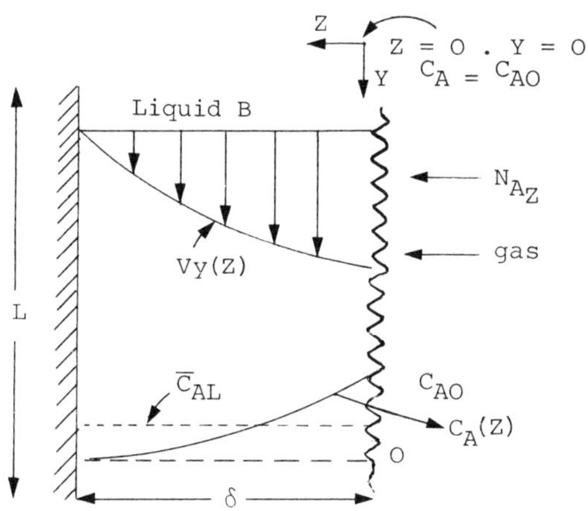

Fig. 13.4

Assumptions:

(1) Mass transfer of A into a falling liquid film.

(2) The diffusion is negligible in the y-direction.

(3) The inlet concentration $C_A = 0$, at steady state.

(4) The gas A is slightly soluble in B (liquid).

(5) The depth of penetration is small compared to the thickness of the film "δ".

General Equation:

$$V_{max} \frac{\partial C_A}{\partial y} = D_{AB} \frac{\partial C_A}{\partial z}$$

Boundary Conditions:

(1) at $y = 0$ $\quad C_A = 0$

(2) at $z = 0$ $\quad C_A = C_{Ao}$

(3) at $Z = \infty$ $\quad C_A = 0$

Solution:

(1) Concentration profile

$$\frac{C_A}{C_{Ao}} = \text{erfc} \frac{Z}{\sqrt{4D_{AB}^{y} / V_{max}}} \tag{13.11}$$

(2) Local mass flux at $Z = 0$

$$N_{AZ}(y)\Big|_{Z=0} = C_{Ao} \sqrt{\frac{D_{AB} V_{max}}{\pi y}} = -D_{AB} \frac{\partial C_A}{\partial z}\Big|_{Z=0} \tag{13.12}$$

(3) Molar flow rate of A across the plane $Z = L$ is

$$N_A = W_A = WLC_{Ao}\sqrt{\frac{4D_{AB} V_{max}}{\pi L}} \tag{13.13}$$

13.6 DIFFUSION AND CHEMICAL REACTION INSIDE A POROUS SPHERICAL CATALYST

The Effectiveness Factor:

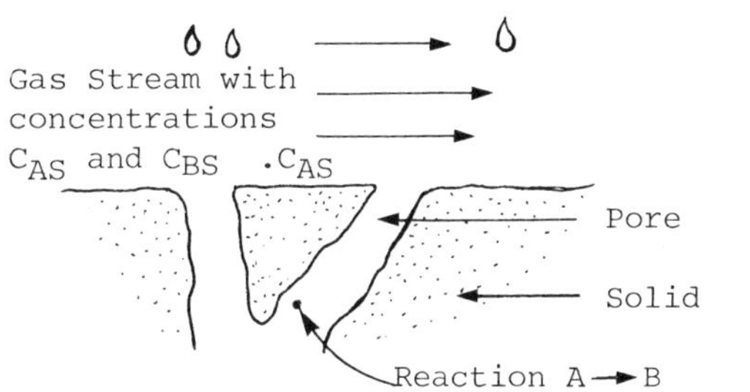

Fig. 13.5

The Governing Differential Equation:

$$\frac{d}{dr}(r^2 N_{Ar}) = r^2 R_A$$

Effective diffusivity is given as below

$$N_{Ar} = -D_A \frac{dC_A}{dr}$$

and rate $R_A = -K''_A \, a \, C_A$

where

a = Surface area per unit volume.

The Boundary Conditions are:

(1) $C_A = C_{AS}$ at $r = R$

(2) $C_A =$ finite at $r = 0$

Solution is given as:

$$\left(\frac{C_A}{C_{AS}}\right) = \left(\frac{R}{r}\right) \frac{\sinh\sqrt{K''_A a/D_A} \cdot r}{\sinh\sqrt{K''_A a/D_A} \cdot R} \tag{13.14}$$

The molar flux at the surface is given as

$$Q_{MS} = 4\pi R\, D_A\, C_{AS} \left\{1 - \sqrt{\frac{K''a}{D_A}} \cdot R \coth\sqrt{\frac{K''_A a}{D_A}} \cdot R\right\} \tag{13.15}$$

Molar rate of conversion

$$Q_{Mo} = (\text{Available surface}) * (\text{Surface reaction rate})$$

or

$$Q_{Mo} = \frac{4}{3}\pi R^3\,(a)(-K_A C_{AS}) \tag{13.16}$$

Effectiveness Factor

$$\eta = \frac{Q_{MS}}{Q_{Mo}} = \frac{3}{K^2}\left(\frac{K}{\tanh K} - 1\right) \tag{13.17}$$

where

$$K = \sqrt{\frac{K_A a}{D_A}} \cdot R$$

For non-spherical particles:

$$R_{NS} = R = 3\left(\frac{V_C}{S_C}\right)$$

where

V_C = Volume of catalyst

S_C = External surface area of single catalyst.

Conversion rate would be:

$$\boxed{|Q_{MS}| = V_P \cdot a \cdot K_A C_{AS}\, \eta_A} \qquad (13.18)$$

where

$$\eta_A = \frac{1}{3\zeta}\left(\frac{3\zeta}{\tanh 3\zeta} - 1\right)$$

and

$$\zeta = \sqrt{\frac{K_A a}{D_A}}\left(\frac{V_P}{S_P}\right)$$

CHAPTER 14

THE EQUATION OF CHANGE FOR MULTICOMPONENT SYSTEM

14.1 EQUATION OF CONTINUITY OF 'A' FOR CONSTANT ρ AND D_{AB}

(1) Rectangular Coordinates:

$$\frac{\partial C_A}{\partial t} + \left(v_x \frac{\partial C_A}{\partial x} + v_y \frac{\partial C_A}{\partial y} + v_z \frac{\partial C_A}{\partial z} \right)$$

$$= D_{AB} \left(\frac{\partial^2 C_A}{\partial x^2} + \frac{\partial^2 C_A}{\partial y^2} + \frac{\partial^2 C_A}{\partial z^2} \right) + R_A$$

(2) Cylindrical Coordinates:

$$\frac{\partial C_A}{\partial t} + \left(v_r \frac{\partial C_A}{\partial r} + v_\theta \frac{1}{r} \frac{\partial C_A}{\partial \theta} + v_z \frac{\partial C_A}{\partial z} \right)$$

$$= D_{AB} \left[\frac{1}{r} \frac{\partial}{\partial r} \left(r \frac{\partial c}{\partial r} \right) + \frac{1}{r^2} \frac{\partial^2 C_A}{\partial \theta^2} + \frac{\partial^2 C_A}{\partial z^2} \right] + R_A$$

(3) Spherical Coordinates:

$$\frac{\partial C_A}{\partial t} + \left(v_r \frac{\partial C_A}{\partial r} + v_\theta \frac{1}{r}\frac{\partial C_A}{\partial \theta} + v_\phi \frac{1}{r \sin\theta}\frac{\partial C_A}{\partial \phi}\right)$$

$$= D_{AB}\left[\frac{1}{r^2}\frac{\partial}{\partial r}\left(r^2 \frac{\partial C_A}{\partial r}\right) + \frac{1}{r^2 \sin\theta}\frac{\partial}{\partial \theta}\left(\sin\theta \frac{\partial C_A}{\partial \theta}\right) + \frac{1}{r^2 \sin^2\theta}\frac{\partial^2 C_A}{\partial \phi}\right] + R_A$$

14.2 SIMULTANEOUS HEAT AND MASS TRANSFER

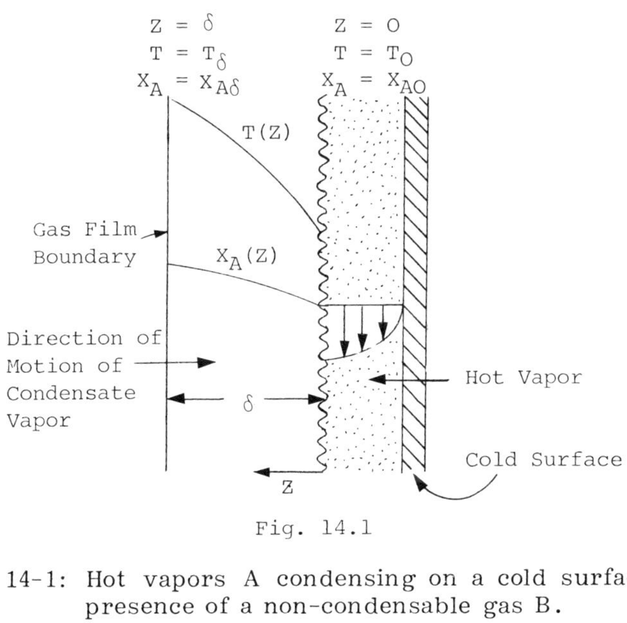

Fig. 14.1

Fig. 14-1: Hot vapors A condensing on a cold surface in the presence of a non-condensable gas B.

Given the system depicted in Figure 14.1 and specifying film boundary quantities at $Z = 0$ and $Z = \delta$, it is desired to determine the concentration profile $X(Z)$ and temperature profile $T(Z)$.

Solution:

Conservation equation for A $\quad \frac{dN_{AZ}}{Z} = 0 \quad$ (14.1)

Continuity of energy $\quad \frac{de_Z}{dZ} = 0 \quad$ (14.2)

Mass flux for diffusion of A through stagnant B is:

$$N_{AZ} = \frac{-c\,D_{AB}}{1 - X_A} \frac{dX_A}{dZ} \tag{14.3}$$

Inserting equation (14.3) into equation (14.1) and integrating yields the concentration profiles

$$\frac{1 - X_A}{1 - X_{Ao}} = \left(\frac{1 - X_{A\delta}}{1 - X_{Ao}} \right)^{\frac{Z}{\delta}} \tag{14.4}$$

Find $\frac{dX_A}{dZ}$ from equation (14.4) and substitute into equation (14.3) and obtain flux:

$$N_{AZ} = \frac{c\,D_{AB}}{\delta} \ln \left(\frac{1 - X_{A\delta}}{1 - X_{Ao}} \right) \tag{14.5}$$

Combining equations (14.4) and (14.5) and rearranging gives

$$\frac{X_A - X_{Ao}}{X_{A\delta} - X_{Ao}} = \frac{1 - \exp[(N_{AZ}/cD_{AB})Z]}{1 - \exp[(N_{AZ}/cD_{AB})\delta]} \tag{14.6}$$

To determine the temperature profile repeat the protocol using the energy flux for an ideal gas, and obtain

$$\frac{T - T_o}{T_\delta - T_o} = \frac{1 - \exp[(N_{AZ}\tilde{C}_{PA}/k)Z]}{1 - \exp[(N_{AZ}\tilde{C}_{PA}/k)\delta]} \tag{14.7}$$

14.3 UNSTEADY-STATE EVAPORATION

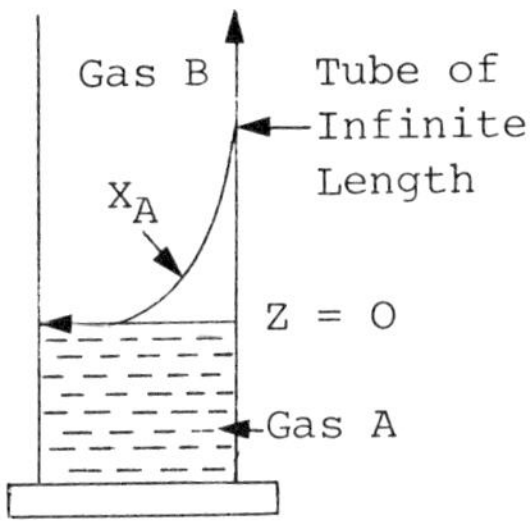

Fig. 14.2

Assumptions:

(1) Temperature and pressure = constant

(2) Molar density C = constant

(3) B is insoluble in A

(4) Liquid level is maintained

General Equation:

$$\frac{\partial c}{\partial t} = - \frac{\partial}{\partial Z} (N_{AZ} + N_{BZ})$$

Initial Condition:

At $t = 0$ $\quad X_A = 0$

Boundary Conditions:

At $Z = 0$ $\quad X_A = X_{Ao}$

At $Z = \infty$ $\quad X_A = 0$

Solution:

(1)

$$X_{Ao} = \frac{\text{vapor pressure of A}}{\text{Total pressure}} = \frac{1}{1 + [\sqrt{\pi}\,(1 + \text{erf}\,\phi)\,\phi \exp \phi^2]^{-1}} \tag{14.8}$$

where

$$\phi(X_{Ao}) = -\frac{1}{2}\,\frac{X_{Ao}}{(1 - X_{Ao})}\,\left.\frac{dx}{dZ}\right|_{Z=0}$$

and

$$X = \frac{X_A}{X_{Ao}}, \qquad Z = \frac{z}{\sqrt{4D_{AB}t}}$$

(2) Rate of vaporization of A from a surface of area S is

$$V_A = S\phi\sqrt{4D_{AB}t} \tag{14.9}$$

14.4 GAS ABSORPTION WITH RAPID CHEMICAL REACTION

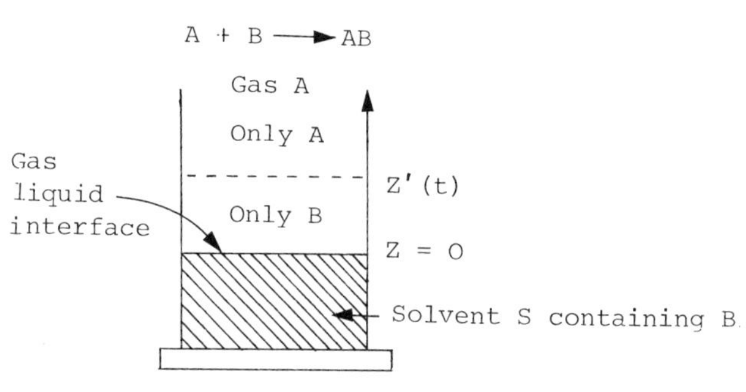

Fig. 14.3

Assumption:

First order irreversible chemical reaction.

General Equation:

$$\frac{\partial C_A}{\partial t} = D_{AS} \frac{\partial^2 C_A}{\partial Z^2} \qquad 0 < Z \leq Z'(t)$$

$$\frac{\partial C_B}{\partial t} = D_{BS} \frac{\partial^2 C_B}{\partial Z^2} \qquad Z'(t) \leq Z < \infty$$

Initial condition: at $t = 0$ $C_B = C_{Bo}$

Boundary Conditions:

(1) at $Z = 0$ $\quad C_A = C_{Ao}$

(2) at $Z = Z'(t)$ $\quad C_A = 0$

(3) at $Z = Z'(t)$ $\quad C_B = 0$

(4) at $Z = Z'(t)$ $\quad -D_{AS} \frac{\partial C_A}{\partial Z} = + D_{BS} \frac{\partial C_B}{\partial z}$

Solution:

(1)
$$\boxed{\frac{C_A}{C_{Ao}} = a_1 + a_2 \operatorname{erf} \frac{Z}{\sqrt{4D_{AS}t}}} \qquad (14.10)$$

(2)
$$\boxed{\frac{C_B}{C_{Bo}} = b_1 + b_2 \operatorname{erf} \frac{Z}{\sqrt{4D_{BS}t}}} \qquad (14.11)$$

where

C_{Ao} = Interface concentration of liquid A

C_{Bo} = Original concentration of B in S

$$a_1 = 1$$

$$a_2 = -\left(\text{erf}\sqrt{\frac{\alpha}{D_{AS}}}\right)$$

$$b_1 = 1 - \left(1 - \text{erf}\sqrt{\frac{\alpha}{D_{BS}}}\right)^{-1}$$

$$b_2 = \left(1 - \text{erf}\sqrt{\frac{\alpha}{D_{BS}}}\right)^{-1}$$

and

$$1 - \text{erf}\sqrt{\frac{\alpha}{D_{BS}}} = \frac{C_{Bo}}{C_{Ao}}\sqrt{\frac{D_{BS}}{D_{AS}}}\ \text{erf}\sqrt{\frac{\alpha}{D_{AS}}}\ \exp\left[\frac{\alpha}{D_{AS}} - \frac{\alpha}{D_{BS}}\right] \tag{14.12}$$

Rate of mass transfer:

$$N_A\Big|_{Z=0} = \frac{C_{Ao}}{\text{erf}\sqrt{\alpha/D_{AS}}}\sqrt{\frac{D_{AS}}{\pi T}} \tag{14.13}$$

14.5 BOUNDARY LAYER THEORY

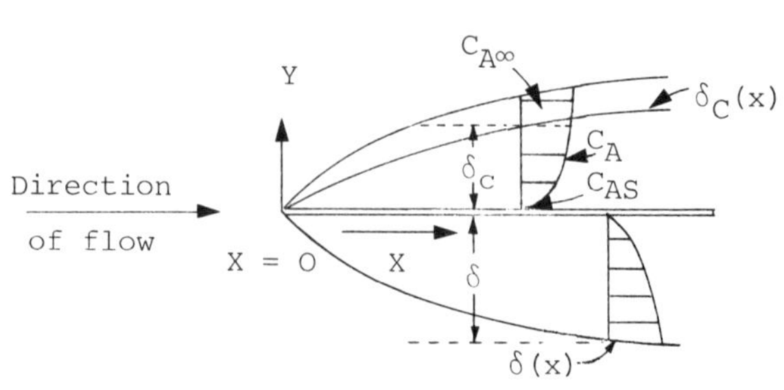

Fig. 14.4

Assumptions:

(1) Time independent flow $\frac{\partial C_A}{\partial t} = 0$

(2) No reaction $\therefore R_A = 0$

(3) $V_Z = 0$

(4) ρ = constant, μ = constant, D_{AB} = constant

Boundary Conditions:

(1) at y = 0 $\frac{V_X}{V_\infty} = 0$

(2) at y = 0 $\frac{T - T_S}{T_\infty - T_S} = 0$

(3) at y = 0 $\frac{C_A - C_{AS}}{C_{A_\infty} - C_{AS}} = 0$

(4) at y = ∞ $\frac{V_X}{V_\infty} = 1$

(5) at y = ∞ $\frac{T - T_S}{T_\infty - T_S} = 1$

Solution:

(1) $\left(\frac{V}{\partial y}\right)_{y=0} = 0.332 \frac{V\infty}{X} N_{Re,x}^{\frac{1}{2}}$ = velocity gradient (14.14)

where $N_{Re,x} = \frac{xV\rho}{\mu}$

(2) Concentration gradient at the (14.15)

surface = $\left(\frac{\partial C_A}{\partial y}\right)_{y=0} = (C_{A_\infty} - C_{AS}) \frac{0.332}{x} N_{Re,x}^{\frac{1}{2}}$

(3) $$\frac{K'_C x}{D_{AB}} = N_{Sh,x} \quad 0.332\, N_{Re,x}^{\frac{1}{2}} \quad \text{Restriction } N_{Sc} = 1 = \frac{\mu}{\rho D_{AB}}$$

(14.16)

(4) $$\frac{\delta}{\delta_c} = N_{Sc}^{\frac{1}{3}} = \left(\frac{\mu}{\rho D_{AB}}\right)^{\frac{1}{3}} \qquad (14.17)$$

(5) $$\frac{K'_C x}{D_{AB}} = N_{Sh,x} = 0.332\, N_{Re,x}^{\frac{1}{2}}\, N_{Sc}^{\frac{1}{3}} \qquad (14.18)$$

(6) Mean mass transfer coefficient K'_C from $x = 0$ to $x = L$ for a width "b" of a plate by integrating as follows

$$\frac{K'_c L}{D_{AB}} = N_{Sh} = 0.664\, N_{Re,L}^{\frac{1}{2}}\, N_{Sc}^{\frac{1}{3}} \qquad (14.19)$$

This shows mean mass transfer coefficient is twice the local mean mass transfer coefficient.

14.6 FILM MASS TRANSFER THEORY

Assumptions:

(1) There is a laminar film next to the boundary.

(2) The molecular diffusion has the same resistance as in the laminar transition and turbulent core region.

General Equation:

$$J_A^* = K'_C (C_{A_1} - C_{A_2}) = \frac{D_{AB}}{\delta_c} (C_{A_1} - C_{A_2}) \tag{14.20}$$

$$K'_C = \frac{D_{AB}}{\delta_c}$$

δ_c = film thickness.